W9-AQX-676

THE
WELL-DECORATED
FLOOR

THE
WELL-DECORATED
FLOOR

beautiful floors and floorcloths

Susan Goans-Driggers

Sterling Publishing Co., Inc.
New York

Prolific Impressions Production Staff:

Editor: Mickey Baskett
Copy: Phyllis Mueller
Graphics: Lampe-Farley Communications, Inc.
Styling: Laney Crisp McClure, Lenos Wallace
Photography: Jeff Herr, Pat Molnar, Jerry Mucklow
Administration: Jim Baskett

Every effort has been made to insure that the information presented is accurate. Since we have no control over physical conditions, individual skills, or chosen tools and products, the publisher disclaims any liability for injuries, losses, untoward results, or any other damages which may result from the use of the information in this book. Thoroughly read the instructions for all products used to complete the projects in this book, paying particular attention to all cautions and warnings shown for that product to ensure their proper and safe use.

No part of this book may be reproduced for commercial purposes in any form without permission by the copyright holder. The written instructions and design patterns in this book are intended for the personal use of the reader and may be reproduced for that purpose only.

Library of Congress Cataloging-in-Publication Data Available

Driggers, Susan Goans.
   The Well-Decorated Floor: beautiful floors and floorcloths / Susan Goans Driggers
      p.cm.
   ISBN 1-4027-0074-1
   1. Painting. 2. Floor coverings. I. Title
  TT385.D75 2000
  645'.1--dc21   00-061893

10 9 8 7 6 5 4 3 2 1

First paperback edition published in 2002 by
Sterling Publishing Company, Inc.
387 Park Avenue South, New York, N.Y. 10016
Produced by Prolific Impressions, Inc.
160 South Candler St., Decatur, GA 30030
© 2000 by Prolific Impressions, Inc.
Originally published in hardcover under the title *Floor Style*
Distributed in Canada by Sterling Publishing
C/o Canadian Manda Group, One Atlantic Avenue, Suite 105
Toronto, Ontario, Canada M6K 3E7
Distributed in Australia by Capricorn Link (Australia) Pty. Ltd.
P.O. Box 704, Windsor, NSW 2756 Australia

Printed in China
All rights reserved
Sterling ISBN 1-4027-0074-1

About the Author
Susan Goans Driggers

Designer and artist, Susan Goans Driggers has made a fascinating career of her love of art and painting. With art classes from the College of Charleston, and an interest in doing her own home decorating such as stenciling a wall border when stenciling materials were not available, Susan began developing her own stencils, tools, and techniques. The talents Susan has acquired are mostly self taught and regarded by her as "a gift from God." Her family has been noted for their ingenuity and entrepreneurship; producing inventors and skilled craftsmen. Creativity seems to be sprinkled throughout their lives.

Susan has created many works of art for some of the most well-known designers in the country. Her work has been displayed in historical homes such as Atlanta's Peacock House, houses in South Carolina's South Battery, and the Mills Mansion Estate in New York's Hudson River Valley. Internationally, her work graces the walls of the privately owned Naworth Castle in Cumbria, England.

While seeking to share her love for art and design with others, she has authored over thirty books and developed technique processes and painting tools that have been patented and distributed in the paint industry and for sale to the general public. Over the past several years she has been featured on many "Do-It-Yourself" type television programs. Her originality has been sought after by some of this country's leading magazines. While putting her "all" into these endeavors, she desires to, as she says, "leave a little part of myself with you." When she teaches her craft, she painstakingly finds the necessary recipe for each of her students that will instill self-confidence, give inspiration, and transfer knowledge. As she reviews her objectives for each project, class or book, her utmost goal is to produce "quality" and "a job well done," making her work a timeless masterpiece. ✍

Author Acknowledgements

As I thought about the care and concern in which this book has been organized and put together, each "thank you" or word of gratefulness did not seem to be enough. Each book I do seems to take on a personality of its own, and this one is "a labor of sharing." From the paint brushes of Donna O'Rourke Mabrey and Patty Cox, to the meal wagon that my sister Patricia always seemed to have warm, ready, and waiting—I received unconditional support. Suzette Faith Goans, my favorite little artist, was always ready and willing to help Aunt Susan with the floor projects. Many homeowners opened up their doors for me while I was executing these beautiful floors. They allowed me to share the vision that a stroke of color can rest upon any surface, even the ones we don't readily think about. I hope the following pages will inspire you to create a masterpiece that will rest beneath your feet.

I give my expressions of love and thankfulness to everyone involved, including Douglas Grant Goans, Patricia & Harold Day, Ruth & David Wray, Linda & Jerry Titus, Mrs. Doris Bray, and Mr. & Mrs. Winford Orr. Your patience has not been overlooked and I do pray that God will richly bless you and your families. ✍

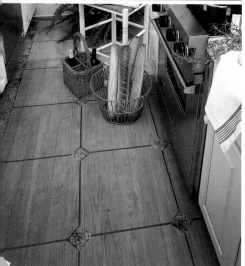

Table of Contents

FLOORS ARE SO NECESSARY AND OMNIPRESENT that it's easy to discount their importance to a decorating scheme. They are a wonderful way to introduce pattern and color to a room without overwhelming it and to add just the right decorative touch to complete or underscore a theme. You can add more lasting decorative elements by working directly on the floor or create a changeable visual feast with rugs and floorcloths. The photographs and projects in this book —more than 25 in all — attest to the decorative value of floors.

The techniques used to decorate floors and floor coverings are time honored ones:

- **Staining**, to enhance the warm tones of wood and let its beauty show and create backgrounds for further decoration
- **Color washing**, for subtle transparency and background effects
- **Stenciling**, to create borders, linear designs, and spot motifs
- **Stamping**, for repeat motifs and random accents
- **Painting**, to create dramatic backdrops and broad expanses of color
- **Faux finishing**, to give surfaces underfoot the look of what they are not—a "granite" inlay and border, tile patterns on concrete, marbleizing on a floorcloth

New products, such as quick-drying acrylic craft paints and stencil gels and acrylic varnishes that resist yellowing make it easier than ever to achieve terrific results. Step-by-step instructions for each project, numerous illustrations and photos, and patterns for making stencils and stamps are included.

Fabulous floors await you!

CHAPTER 1

FLOORS & FLOOR COVERINGS

Almost any kind of floorcovering can be decorated, except ceramic tile and wall-to-wall carpeting. Following are some guidelines for the types of floors that have been decorated for this book.

Wood Floors

Wood floors can be **hardwood** (most commonly oak or maple) or **softwood** (pine or fir). Generally speaking, hardwoods are more durable and more expensive than softwoods.

The boards of wood floors can be cut into simple planks or milled for a tongue-and-groove fit. These floors can be stained, bleached, stenciled, stamped, and painted, and are especially well-suited for techniques like staining, where the wood grain will be visible.

Sheets of **plywood**, usually 4' x 8' in size, may be found on the floors under wall-to-wall carpeting or vinyl tile or in converted attics. Plywood floors are best painted and decorated with grid designs which disguise the seams in the floor where the sheets meet.

Concrete

Concrete floors can be found under carpeting in houses with slab construction and in basements, garages, and carports or on patios. A concrete floor can be decorated without a painted basecoat—the natural or weathered color of the concrete can be used as the base and with decorative painting on top of it.

Concrete stepping stones are also a possibility for decorating. They can be found ready-made in several sizes that are ready to decorate. Stepping stones can be decorated and used indoors and out.

Vinyl, Linoleum, and Asphalt Tile Flooring

These floors can be painted and designed if you follow precisely the instructions for preparing the surface. They will not hold up as well as wood or concrete in areas that receive heavy traffic or lots of wear and tear.

Floor Coverings

Canvas Floorcloths

Traditional floorcloths are made from canvas fabric. Canvas is available in varying thread counts; use a finer weave canvas for high traffic areas. For best results, purchase 100% cotton canvas. You can buy artist's canvas at art supply stores, cotton canvas at fabric stores and crafts supply stores, and heavy woven canvas at tent, awning, and boating equipment shops. You will save yourself many problems by purchasing pre-primed canvas or "floorcloth canvas" that is ready to decorate.

Basic Steps for Making a Canvas Floorcloth

1. Cut canvas to desired size plus 1" hem allowance on all sides.
2. Turn hem edges under 1" and press with an iron. Cut a triangle off of each corner to make a mitered edge.
3. Apply fabric glue along one hem edge. Adhere to cloth back. Press with iron.
4. Apply gesso or neutral glazing medium to entire surface of floorcloth back. Let dry. Apply gesso or neutral glazing medium to entire surface of floorcloth front. Let dry.
5. Paint, stamp, and stencil designs with acrylic paints, crackle mediums, colored paint glazes, and stencil paints. Let dry completely.
6. Apply two to four coats gloss finish to floorcloth surface, allowing each coat to dry between applications. Finish with one coat satin finish on floorcloth front.
7. To keep floorcloths from sliding, purchase a foam rug pad, or coat the back with a skid-proof rug backing.

Note: Floorcloths may be ironed between all stages of painting when the paint is dry and even after all finishing coats have been applied. **Do not** place iron directly on the finish; use an old bed sheet as a pressing cloth.

To store a floorcloth, roll it up gently. **Don't** fold, bend, or crease it.

Floorcloths Made from Vinyl Flooring

Floorcloths also can be painted, stamped, and stenciled on the smooth **back side** of vinyl flooring. Vinyl flooring can be found in remnants at many home improvements stores. Vinyl floor is inexpensive, easy to cut any size, and easy to decorate. It will not slide as much as a canvas floorcloth. A thin foam rug pad can be used under it if you desire.

Sisal Rugs or Carpets

Sisal is a woven natural fiber. It accepts paint well, soaking it up almost like a sponge. Be sure to seal it so dirt or stains won't ruin the surface. I usually spray the completed rug with a rug-grade stain repellent.

You can purchase sisal either as area rugs (bound with cloth or invisibly bound and latex-backed for a non-slip rug) or as wall-to-wall carpet, which you can have cut and bound to size. Be sure to choose a finely woven sisal.

Woven Rag Rugs

Rag rugs usually are made from cotton fiber, but some are made from synthetics. You will need to test your rug on the back to see how it accepts paint. (Most I have tested have worked well.) I usually add one or two drops of vinegar to each teaspoon of paint that I use for decorating. The vinegar helps the paint to adhere and set into the fabric better, making the paint less apt to fade. To protect the painted design, spray the surface with a fabric stain repellant after the paint has dried thoroughly and after each washing to prevent stains and dirt from damaging the surface.

Other Floor Coverings

Woven straw mats and coir mats are other surfaces that can be decorated with paint. ✍

Canvas floorcloth

Vinyl flooring floorcloth

Sisal rug floorcloth

CHAPTER 2

FLOOR DECORATING SUPPLIES

Finding the paints, finishes, and other supplies for decorating your floor is not as daunting as you might think. Most home improvement, hardware, or even craft shops will have the items you need to create beautiful floors and floorcoverings.

Paints & Stains

A variety of paints and stains can be used on interior or exterior floors. Some require a protective sealer or finish so they will withstand foot traffic and resist soiling, while others are formulated to hold up under normal traffic without a sealer. The paint department at your local hardware store or home improvement center and paint stores are excellent sources for information on the proper products to use on your floor. For best results, always follow the manufacturer's instructions and recommendations regarding preparation, ventilation, drying times, re-coating intervals, and cleanup.

Paints and stains are either **oil-base** or **waterbase**. I usually recommend waterbase paint because it cleans up with soap and water and has little or no odor, but oil-base products can and should be used in some cases.

Stain should be used for floors and decorating when you want the wood grain or other base to show. You can create wonderful "faux parquet" flooring by using stains of various shades.

Paint should be used when opaque, solid coverage is desired. Most paints are formulated to cover the surface with one to two coats. Paint can be used to create a film of color for color staining by diluting the paint with water or mixing it with a **neutral glazing medium**, but will still have an opaque appearance. If you use paint to create a colored stain, you have to work quickly and very carefully to avoid overlap marks in the color.

Exterior paints and stains are usually formulated to resist mildew and damage from ultra-violet rays. Products intended for exterior use are less affected by water and sun, heat and cold and resist fading and peeling. They usually don't require a protective sealer. If, however, you wish to seal the surface, always use a product intended for exterior use.

Interior paints and stains are made for surfaces that are protected from weather and direct sun. You can use exterior paints and stains for inside projects (unless the label directs otherwise), but I do not recommend interior paints and stains for exterior projects—they are just not as durable.

Gesso or **neutral waterbase glazing medium**, is needed for priming canvas before painting a canvas floorcloth. ⊚

Finishes

The projects in this book are intended to be finished with clear sealers when sealing is necessary. Options include oil-base and waterbase varnishes, oil-base and waterbase polyurethane, polyacrylic finishes, acrylic varnish, paste wax, and tung oil. Finishes also have different sheens when they dry: gloss (which looks hard and shiny), satin (which has a softer sheen), and matte, a flat finish with a slight sheen and no shine.

Choose a sealer according to your particular requirements that will be compatible with your floor: whether the floor is interior or exterior, what types of products were used on the floor (oil-base or waterbase), whether the floor will receive light or heavy traffic, how often you want to re-apply the finish (wax must be applied more often than polyurethane, for example), how you want to clean the floor (e.g., oil floors should be dusted, but many varnished floors can be mopped), whether your household includes pets or children. Discuss your requirements with someone in the paint department of your local hardware store or home improvement center or at a paint store, and read the labels of recommended products.

Remember:
- Never use waterbase paint or stain over oil-base paint.
- You can use waterbase paint or waterbase stain over oil-base or waterbase stain, as long as it does not have an oil-based varnish or sealer already over it.
- Never use oil-base paint or stain to create decorative patterns on top of an oil-base stain if you are using an oil-base finish. If you do, the top coat finish will cause the decorative patterns to smear when it is applied.
- It is essential that you use water-based varnish if you have painted designs with oil base paint. Reason: oil will remoisten, causing smearing. Water-based varnish will not remoisten the oil paint under it.

Brushes & Applicators

Most tools for applying paints and stains, such as brushes and rollers, work with both waterbase and oil-base products. Buy applicators appropriate to the product you're applying and use them as the manufacturer intended. Paint and hardware stores and home improvement centers carry a dazzling array of **brushes and rollers** in a variety of sizes and **specialized applicators** such as paint pads and mitts. My best general advice is to buy quality tools (this applies especially to paint brushes) and clean them carefully, following the manufacturer's recommendations, after each use.

Stenciling can be done with **stencil brushes, bristle paint brushes, sponges, or sponge applicators.** You may also want to experiment with using an **airbrush** for stenciling—there are many different types on the market that work well and hold very little paint, so they are perfect for small projects.

Marking & Measuring Tools

You can use **chalk pencils, lead pencils** (they're not really lead anymore, they're made of carbon), or a **chalk line** to mark your floor. Whatever you use, use it lightly. Choose a chalk color that will blend with your floor and use a damp rag to wipe away as much of the residue as you can. Erase all pencil marks. If you don't remove your marks, they will show through your stain (and many paints), spoiling the appearance of your finished floor.

A **wash-away fabric marker** is a good tool for marking canvas floorcloths.

You'll need a **measuring tape** (a 25-foot one is long enough for most rooms) and/or a yardstick for measuring.

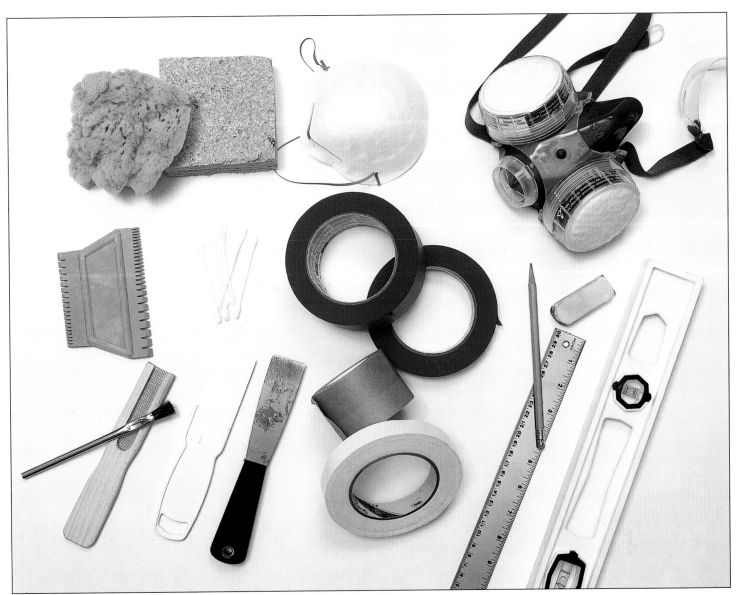

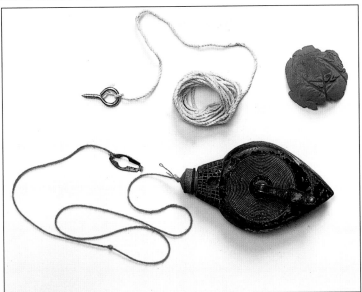

Chalk line and plumb line

Supplies for Stenciling

Stenciling is used to create repeated borders and designs on surfaces. A huge variety of **pre-cut stencil designs** are available or you can cut your own from **stencil blank material. Acrylic craft paints** or **stencil paints or gels** are most often used for stenciling.

To stencil, secure the stencil to the surface with **masking tape**. Apply paint through the openings of the stencil, using very little paint on a **stencil brush, sponge, or foam applicator**. When a section is complete, carefully lift the stencil, move it to the next area (pre-cut stencils include alignment marks for positioning), and continue to apply paint. When properly cleaned after each use, stencils can be used again and again.

To cut your own stencil, you will need a **craft knife or a stencil cutter,** a cutting surface such as a piece of thick glass with smoothed or taped edges, to use as a cutting surface or a **cutting mat** and **stencil blank material.** The stencil blank plastic material is usually sold where ready-cut stencils are found. It makes stenciling easier if a stencil overlay is cut for each color that is

going to be stenciled. When doing this be sure to make alignment marks on each stencil overlay by drawing in with permanent ink, areas that will not be cut. For example, when stenciling a flower, you would have a stencil overlay for the petals, with the petals cut out and the leaves drawn in with ink. Then there would be another stencil overlay with the leaves cut out and the flowers drawn in with permanent ink. It is also possible to cut just one stencil of the entire design, then as you are stenciling one area, tape over the areas that you are not stenciling at the time. This will keep from getting unwanted paint in an area.

Supplies for Stamping

Stamping is a quick and easy way to add design motifs to surfaces. You apply paint to the face of the stamp, press it to the surface, and lift. Stamping is also called "block printing."

Pre-cut stamps in a variety of shapes and sizes are available at crafts stores and home improvement centers. They are usually labeled "**design stamps**" or "**printing blocks**." You also can cut your own stamps from **cellulose sponges**, from mousepad material, from soda can hugger material, or **blank stamping material** that can be found where stamping supplies are sold. When transferring the design to the pad, use a tracing wheel and waxy dressmaker's carbon paper. Cut out the designs with a craft knife.

Computer mouse pads are great for making stamps, with a few cautions:

- Do not use pads that have a crinkled pattern on the rubber—if you can only find this type, use the cloth side of the pad for stamping.
- Do not use areas that have letters or designs—they will show up on your stamped designs

Natural materials, such as leaves and vegetables, also may be used to make patterns from, or they can be used as actual stamps.

Supplies for Masking Tape Bands & Stripes

Use **low tack masking tape** for marking grids, bands, and stripes. It's easy to remove and leaves little residue.

When marking narrow bands (1/2" or 1/4" wide, for example), it's always faster and more accurate to use a piece of masking tape than to measure and try to mark narrow widths with a chalk line. Here's how:

1. I measure my first line and mark it with a chalk line.
2. I place masking tape (in the width I want the band to be) along the line—this gives a straight line without waves or wiggles.
3. I position a wider piece of masking tape on each side of the tape and carefully remove the center tape, leaving a space just the right size.✐

CHAPTER 3

FLOOR PREPARATION

New Interior Wood Floors

New (raw) wood floors should be sanded only if needed to smooth them. Previously finished wood floors should be sanded smooth with a professional floor sanding machine to expose raw wood. Sanding machines can be rented from most tool rental businesses. Or you may hire professional wood floor finishers. If you hire someone, be sure they understand you will need time to decorate the floor after they sand it and before they apply the finish.

Basic Steps for Decorating Interior Wood Floors:

1. Sand the floor (this is often called "heavy" or "rough" sanding) to expose raw wood. Be sure to protect adjacent rooms from dust and wear a mask if you're doing the sanding yourself.
2. Remove all dust by vacuuming with a vacuum cleaner.
3. Sand again with a finer grit of sandpaper to smooth the wood.
4. Remove all dust by vacuuming with a vacuum cleaner.
5. Apply base stain or paint. (optional)
6. Decorate the floor.
7. Apply a light coat of finish to seal the floor. Allow to dry thoroughly, 24-48 hours.
8. Lightly sand with a very fine grade of sandpaper to smooth and buff the floor. **Caution:** Be careful! If your sandpaper is too rough or if you sand too long in one spot, you can damage your design.
9. Vacuum the floor to remove dust.
10. Apply additional coats of finish. Allow to dry between coats.

Staining a Previously Painted Floor

Sanding should remove the paint unless it has soaked into the grain. If that's the case, design the floor so the paint color works for you, not against you. Here are some options:

Option 1: Apply a color wash to the floor instead of a wood-tone stain. If the floor was painted gray or blue or white, use a white or off-white wash. The floor will have a "pickled" or "milk washed" look that will hide the old paint.

Option 2: Have a painted base. After sanding, re-paint the floor with the color of your choice. Choose a paint suitable for floors.

Option 3: Choose a dark stain to lessen the contrast between the color of the floor and a dark paint color.

Painting a Previously Painted Floor

If you wish to re-paint a previously painted floor, you'll need to prepare it properly and choose the right kind of paint.

Basic Steps for a Painted Floor:

1. Scrub the floor to remove dirt, wax, and grease. Let dry. Remove any loose paint by scraping and/or sanding the surface, and vacuum up any dust. **CAUTION! Old paint can contain lead, so be sure to wear protective clothing and a mask.**
2. Determine what type of paint is on the floor by using a solvent in an inconspicuous spot. (Ask for help at the paint store when choosing a solvent.)
   • If paint is oil-base, an oil-base paint must be used to re-paint and decorate the floor. *Exception:* If old oil-base paint is weathered, aged, or more than a few years old, you can use rough sandpaper with an electric sander to etch or score the surface, making sure that you thoroughly work every inch of the floor. Then you can paint the floor with waterbase paint. If the floor is not properly sanded, the waterbase paint will wear away quickly or even peel off the surface.
   • If paint is waterbase, you can re-paint with either a waterbase or oil-base paint.
3. Paint the floor, following the paint manufacturer's instructions.

Wood Floors with Glued Floor Coverings

You'll need to scrape off the floor covering to reveal the wood, removing as much of the glue residue as possible and being careful not to damage the underlying wood. You can buy a floor scraper at a hardware store. Tool rental outlets have a machine that will do the job—the machine may be the best option if the flooring is tightly adhered. You'll need to sand to remove all traces of remaining glue residue.

Damaged Floors

Old wood floors may have rotten, broken, badly stained, or otherwise damaged areas. In that case, you'll need to replace the damaged wood with boards of the same type and thickness. To make the new wood blend with the old, study the old wood and note what kinds of holes, scratches, or wear are present. You can "distress" the new wood to match the old with a hammer or screwdriver and stain the new wood so it will have the same appearance. I recommend buying many shades of stains so you can test and work different colors until you find a combination that matches or blends with the old wood.

Old wood floors may have nail holes from carpet or linoleum that was tacked down along the edges or in front of the doors. These holes usually can't be completely filled or sanded so they become invisible, so I design a floor treatment that camouflages the problem. One solution is to create a painted border or faux finish along the edge of the floor, such as faux granite or marbleizing. It's always easier to blend color tones on smaller spaces than on large running areas.

Vinyl, Linoleum, and Asphalt Tile Flooring

Remove all wax, soil, and grease. Sand the surface to etch it so the paint will adhere properly. You need to sand the surface enough to etch it but not so deeply that you damage the flooring. ALWAYS wear a protective ventilator when sanding!

Many of these floors have raised or indented patterns that will be visible after sanding. You cannot sand the patterns from the surface; you simply have to choose a design that works with the pattern and incorporate the pattern into your design.

Some paints adhere better to plastic or rubber materials; check with your local paint or specialty shop for advice and buy the appropriate products.

Exterior Wood Decks and Concrete Patios

I always have exterior floors cleaned before I paint them—proper cleaning helps the paint adhere properly, so it lasts longer. Be sure to clean all dirt, grease, and stains from the floor. It's especially important to remove mildew—if you don't, it can continue to grow, even under the new paint. Use a cleaner sold for this purpose, and be sure to read and follow the manufacturer's instructions.

A pressure washer is often used for a thorough cleaning—you can rent a machine or hire a professional who has one. Be sure the floor is rinsed thoroughly and allowed to dry completely. The time needed for drying depends on the heat and humidity. Any moisture left in the wood or concrete may cause the paint to not harden properly and the durability may suffer.

Interior Concrete Floors

Interior concrete floors, such as enclosed porches, basements, or garages, should be free of dirt, stains, and mildew. Scrub with a brush and use cleansers formulated to deal with what's on the floor. Rinse thoroughly to remove any residue. Allow to dry. ✍

CHAPTER 4
DESIGN PLACEMENT

The sizes of designs and the widths of borders depend on the size of the room, the furnishings, and other patterns in the decor. Doorways and thresholds should be included in your room design. If you have a raised threshold, you'll need to decide which room the threshold should match.

It's a good idea to make proofs of your design by drawing or tracing them on paper, posterboard, or kraft paper. Then position the proofs on the floor and study the effect. Enlarge or decrease the size and vary the placement until it looks right to you. After you've determined where the design will be placed, use very light pencil markings (you'll have to erase them later) or a chalk pencil or a chalk line to mark the design placement on the floor. Choose a chalk color that harmonizes with the floor color—it will be easier to remove.

Overall Designs

An overall design covers the entire surface of the floor—repeated stripes, diamond shapes, and blocks and square, rectangular, circular, and octagonal designs can be considered overall patterns. Overall designs generally allow for areas of the floor to show through.

To determine placement of overall designs, first measure the floor space from corner to corner to find the center of the floor. Starting at the center ensures that motifs will end up the same against each wall and will not appear lopsided. Then divide the floor into sections of uniform size to create a grid for placing the shapes.

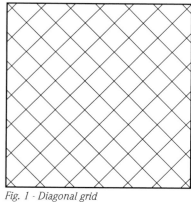

Fig. 1 - Diagonal grid

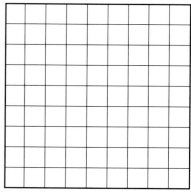

Fig. 2 - Square grid

An overall design — painted stripes cover the entire floor.

Grids

Grids also are used to create a geometric pattern. They can be used alone or as backgrounds for flowers, leaves, or other motifs. To create a grid, section off the floor as you would for an overall pattern. See Fig. 1 and Fig. 2

This grid includes faux granite inlays where the grid lines intersect.

Checkerboards

Checkerboard patterns can be used to create an overall background, to create a grid background for another design, or to add stained or painted sections to the floor.

To create a checkerboard, follow the directions for overall patterns and mark the floor with squares of equal size. Paint or stain, alternating two colors.

This floor is marked into squares and stained with two alternating colors.

Borders

Borders can be bands of color, linear running designs, a combination of the two, or motifs joined by bands or linear designs. The border will seem less structured without bands. When laying out border design, follow these steps to ensure that the border will appear to flow evenly from corner to corner:

1. Measure the distance from corner to corner along the wall. Divide in half to locate the center of the wall's floor space.
2. Start in the center and work out to each side so the pattern will end up the same at each corner.

   • *Consideration:* You may need to "flip" the design to create a mirror image. See Fig. 3.

   • *Consideration:* If a wall has a fireplace hearth or an area that juts out into the room, then start the pattern in the center of that area and work left and right from that point. See Fig. 4.

A border without bands.

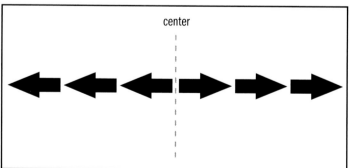

Fig. 3 - Border design using a mirror image.

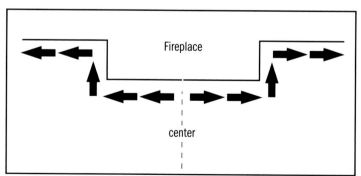

Fig. 4 - Placing a border around a fireplace hearth.

Randomly Placed Designs

Random placement is simply scattering motifs (leaves, flowers, geometric designs, etc.) over the surface. It's not really random—you plan it to look that way. I usually make samples of the designs on paper or poster board and position them on the floor to judge where the motifs should be placed and how many should be applied. An easy way to do this is to use a photocopier to make multiple copies of the design.

Medallions

Medallion designs are round or square motifs placed in the center of the room or at a focal point such as a fireplace, entrance, or window. (Fig. 5) They also can be smaller round or square motifs placed in a repeated pattern. (Fig. 6) Medallion designs differ from random placement designs in that they are meant to look planned and intentionally placed.

The placement of the furniture on this porch was taken into account when planning the random placement of these floral motifs.

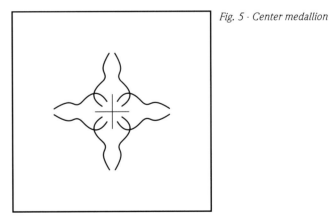

Fig. 5 - Center medallion

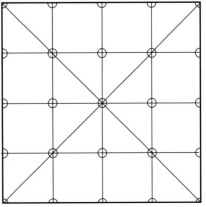

Fig. 6 - Medallion motifs placed on a grid

CHAPTER 5

FLOOR DECORATING PROJECTS

The projects in this book show a variety of ways to decorate and embellish all kinds of floors, indoors and outdoors, with painting, stenciling, staining, and stamping. You'll find ideas for decorating stained wood floors to look like parquet or inlay; how to create a great looking patio from cement stepping stone squares; ways to dress up a deck floor or a screened porch; and how to make stairs look great.

There are also ideas for turning plain rugs and floorcloths into one-of-a-kind works of art. Floorcloths of all types are featured — from quick and easy ones to some so gorgeous that they will be a focal point in a living room or a dining room.

Step-by-step instructions and numerous photos have been provided to inspire and instruct you as you create your masterpieces underfoot.

Patterns for the floor designs can be found at the end of the project chapter. These patterns can be used to create your own decorated floors.

Pictured at right: Hexagons & Diamonds

Hexagons & Diamonds
Stenciled "Faux Parquet" Border on Wood Floor

*With simple stenciling, it's possible to give the look of an elegant parquet border to any floor. This simple border is a continuous design punctuated by bands of color. This design was done on a **newly sanded, raw** wood floor. It was sanded, stenciled, and varnished, but not stained. However, it can be done in darker colors on an existing stained wood floor.*

The pattern for this floor is at the end of the projects chapter.

1 • Supplies

Stains & Finishes:
Waterbase stains: mahogany, walnut, pecan
Finish of your choice

Tools & Equipment:
Masking tape: 1/4", 1/2", and 1"
Chalk line
Stencil blank material
3 stencil brushes
Tracing paper and pencil
Cutting mat
Craft knife
Permanent fine tip marker

Stain Colors Used

mahogany

pecan

walnut

2 • Preparation

If working on a new, raw wood floor, hire a professional to sand the floor or do it yourself with a floor sander. Vacuum up dust.

3 • Making the Stencil

1. Place tracing paper over the stencil pattern found at the back of this book. Trace the pattern on tracing paper with pencil.
2. Enlarge design on a copy machine if needed to fit the proportion of your room. This border is 9" wide—the stencil design is 5" wide, the wide bands are 1", and the narrow bands are 1/2" and 1/4".
3. Trace the stencil patterns on stencil blank material with a fine tip marker.
4. Place the stencil blank material on the glass. Using a craft knife or stencil cutter, carefully cut out the designs. The stencil can be cut into three overlays, one for each color of stain; or cut as one stencil, with areas taped off when stencil colors change. The wide bands of hexagons and the diamond in the center of each hexagon are pecan; the small bands of each hexagon are walnut; and the diamonds between each hexagon are mahogany.

4 • Painting the Wide Bands

Use masking tape to create the bands.
1. Measure in from the base molding 3-1/2" at intervals all around the floor. Mark a straight

Photo 1. Before photo showing floor before the finish and stain was sanded off.

Photo 2. Stenciling the design.

line at this mark with a chalk line all around the border edge. Position 1" tape along the line, placing it on the side of the line farthest in from the wall. Position masking tape on each side of the first piece of tape. Remove the first piece. This will give you a 1" band that is straight.

2. Measure in from the base molding 9". Mark with a chalk line. Position 1" tape along the inside of the line. Position masking tape on each side of the first piece of tape. Remove the first piece. This will result in another 1" wide band.

3. Apply walnut stain to the wide bands. Remove tape. Let dry completely.

5 • Stenciling

Position the stencil between the wide bands, centering the design between the bands, 1/2" from each. Stencil the design with the three stain colors. Let dry.

6 • Applying the Narrow Bands

1. Tape off a band 1/2" wide on the side of the wide band nearest the wall. Apply mahogany stain to the band. Remove tape. Let dry.

2. The last band is positioned 1/4" from the inner 1" band and is 1/4" wide. (See photo for placement.) Apply tape and then stain with mahogany stain. Remove tape and let dry.

7 • Finishing

Seal the floor. ✍

Photo 3. Masking off the 1/4" inner border.

Leaf Garland
Stenciled Leaves on a Stained Wood Floor

Leaves are stenciled on a stained oak floor to form an overall design that is placed on a diagonal grid. There is one stencil overlay for each of the two colors used.

A pattern for the design is at the end of the projects chapter.

1 • Supplies

Stains, Paints & Finishes:
Oil-base wood stain: pecan color
Waterbase stencil gels: forest green, brown
Finish of your choice

Tools & Equipment:
Brushes and rags for applying stain
Chalk line
Measuring tape
Stencil blank material
2 stencil brushes
Tracing paper and pencil
Cutting mat
Craft knife
Permanent fine tip marker

Stencil Gel Colors Used:

brown

forest green

2 • Preparation

1. Hire a professional to sand the floor or do it yourself with a floor sander. Vacuum up dust.
2. Stain the wood floor, following the stain manufacturer's instructions. Let dry.
3. Measure and mark a diagonal grid on the floor. Use a chalk line to mark the stencil placement. (**photo 1 on page 30**)

3 • Making the Stencils

1. Place tracing paper over the stencil patterns found at the end of the projects section in this book. Trace the patterns on tracing paper with pencil. Enlarge as needed.
2. Trace the stencil patterns on stencil blank material with a fine tip marker.
3. Using a craft knife or stencil cutter, carefully cut out the designs. A stencil overlay is cut for the leaves and another overlay is cut for the darker stem/vein area.

4 • Stenciling

1. Position the leaf stencil overlay along the chalk lines and tape in place. Stencil the leaves with forest green, re-positioning the stencil along the floor as you work. Stencil in one direction, then the other. (**photo 2**)
2. Position the second overlay and tape in place. Stencil the second overlay with brown, re-positioning the stencil along the floor as you work. Stencil in one direction, then the other. (**photo 3**) Let dry.

5 • Finishing

1. Wipe away the chalk lines with a damp rag. *Hint:* if a water dampened cloth does not remove the chalk line powder, use a cloth dampened with mineral spirits. Do not make the cloth too moist and do do not drip any on the floor. (**photo 4**)
2. Seal the floor. ✍

Photo 1. Marking the grid with a chalk line.

Photo 2. Stenciling the leaves with forest green stencil gel.

Photo 3. Stenciling the second overlay with brown stencil gel.

Photo 4. Wiping away the chalk dust with a damp rag.

Fabulous Fern
Stencil Resist on a Stained Wood Floor

In this breakfast room, fern motifs are randomly placed on a pine floor. I wanted the fern designs to appear lighter than the background (sort of like a fossil), so I stenciled them with clear varnish (this is called "creating a resist") on the sanded floor before applying the darker stain.

This project was done on an old pine floor that had been painted many times. Paint had soaked into the grain of the wood and was impossible to remove completely. Because the floor color was uneven, I chose a deep, rich colored stain to blend the wood tones for a unified look.

Patterns are provided for the fern motifs at the end of the projects chapter, but a pre-cut stencil also could be used.

1 • Supplies

Stains, Paints & Finishes:
Oil-base wood stain: warm walnut
Clear waterbase varnish

Tools & Equipment:
Brushes and rags for applying stain
Stencil blank material
Stencil brush
Tracing paper and pencil
Cutting mat
Craft knife
Permanent fine tip marker

2 • Preparation

1. If starting with an old wood floor, hire a professional to sand the floor or do it yourself with a floor sander. Vacuum up dust.
2. Place tracing paper over the stencil patterns in this book. Trace the patterns on tracing paper with pencil.
3. Enlarge designs as needed and make paper proofs of the patterns on a copy machine.
4. Position the proofs on the floor to determine the placement of the designs.

3 • Making the Stencils

1. Trace the stencil patterns on stencil blank material with a fine tip marker.
2. Place the stencil blank material on the glass. Using a craft knife, carefully cut out the designs.

4 • Stenciling

Stencil each pattern with clear waterbase varnish, using a stippling motion with a light touch (an up-and-down pouncing motion) and an almost dry brush. After you have completed one motif, let it dry and stipple it again with varnish.

Tips:
- If you don't start with an almost dry brush, the thin liquid varnish could seep under the openings of the stencil. If it does, you won't get a crisp, clean edge.
- It is best to build the solid layer of varnish within the cut-out openings of the stencil, applying two (or even three) coats to ensure solid coverage.
- If the varnish does not solidly cover the stencil design, it will not create a clear image when you apply the stain.

Continue and complete the random application of the stenciled designs. Allow to dry thoroughly.

continued on page 34

continued from page 32

5 • Staining

Stain the wood floor, following the stain manufacturer's instructions. Apply the stain the same way you would if there were no stenciling on the floor—work with the grain of the wood along the boards. I used a rag to apply the stain. When you rub over the fern motifs, you will be able to instantly see the designs. As you work, notice how the stenciling looks—it looks the same wet as when it is dry. You can work with the stain and the pattern as you go to achieve the look that you desire. If the designs are not clear in some areas, wrap a clean dry cloth tightly around your index finger (so you can see what you're doing) and, before the stain dries, wipe some of the stain off the leaf shapes. (This happened on my floor in areas where the wood seemed to be extra hard and the varnish did not soak into the wood as it had in more porous areas.) Allow to dry 24-48 hours.

6 • Finishing

Seal the floor with waterbase varnish.

Photo 1. Before the floor was sanded. It had many layers of paint.

Photo 2. Stenciling with varnish to create a resist. You can see how uneven the floor color is before staining.

Photo 3. Stenciling a different motif with varnish. The first stenciled design is barely visible.

Photo 4. Rubbing stain over the floor. The stain doesn't penetrate the stenciled designs—they resist it.

Twining Vines
Stenciling on Stained Wooden Stair Risers

There is a funny story behind the green stair color. This was a new construction and the owner was giving the "floor man" instructions as to what color of stain went where. The downstairs floors were to be a light fruit wood and were to include the stairs. The beautiful green color was picked out for some bookcases upstairs. The owner said "the green is for upstairs." So that is what the floor man did — he stained up the stairs with green. But this was a mistake that turned into something beautiful. The stairs are a striking focal point that lead up to a landing that has a very large window that overlooks a hardwood tree-covered valley. So in the spring of the year the green of the stairs matches the beautiful new leaf growth seen through the landing window. The twining vine stencil only adds to the rustic beauty.

The pattern is included on the next page.

1 • Supplies

Stains, Paints & Finishes:
Oil-base green stain
Waterbase stencil paint: light gray
Finish of your choice

Tools & Equipment:
Brushes and rags for applying stain
Stencil blank material
Soft, flat bristle brush for stenciling
Tracing paper and pencil
Cutting mat
Craft knife
Permanent fine tip marker

2 • Preparation

1. Prepare the floor for staining. (These stairs were newly installed and only required minimal sanding and vacuuming.)
2. Stain the treads and risers with green stain. Let dry.

3 • Making the Stencil

1. Place tracing paper over the stencil pattern in this book. Trace the pattern on tracing paper with pencil.
2. Enlarge design on a copy machine as needed to fit your stair risers.
3. Trace the stencil patterns on stencil blank material with a fine tip marker.
4. Using a craft knife or stencil cutter, carefully cut out the designs.

4 • Stenciling

Stencil the design, using a soft flat brush. Stroke the color on the surface unevenly, creating a primitive appearance. Let dry.

5 • Finishing

Seal the surface.☺

Pattern for
Twining Vines

Instructions on page 36
Enlarge to 355% for actual size

Tulips & Squares
Stamped & Washed Wood Floor

This new pine floor has been divided into checkerboard squares and color washed with sunny lime and tropical turquoise glazes. After the painting of the checkerboard was completed the floor was sanded for a homey distressed look. Colorful tulips are stamped randomly and very lightly sanded when dry. Rather than applying a varnish, I chose to wax this floor. In time, it will allow it to get more worn.

The patterns for this floor are at the end of the projects chapter.

1 • Supplies

Stains, Paints & Finishes:

Neutral waterbase glazing medium

Acrylic paints (for color washing): light turquoise, avocado, *optional* — gray white

Acrylic craft paints (for stamping): white, sunflower yellow, lavender, pink, medium green, dark green

Floor wax

Tools & Equipment:

Cellulose sponge

Containers for mixing glazes

Chalk line

Masking tape, low tack

Fine and medium grit sandpaper

Stamping material (you can use blank stamping material from a craft department, mousepad material, or can cooler material)

Craft knife

Tracing paper and pencil

Carbon transfer paper, white

Dressmaker's tracing wheel (optional)

#12 artist's flat brush

Round artist's brush

continued on page 42

Paint Colors Used:

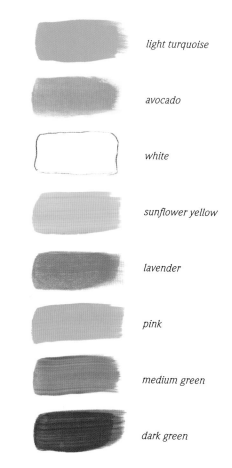

light turquoise

avocado

white

sunflower yellow

lavender

pink

medium green

dark green

continued from page 40

2 • Preparation

1. Prepare the floor. (This floor was newly installed and required only minimal sanding and vacuuming.)
2. Measure and mark the checkerboard squares with a chalk line.
3. Tape off alternating squares with low tack masking tape. Press the tape firmly to the floor.

3 • Color Washing the Squares

1. In one container, mix one part avocado paint and one part glazing medium. For each cup of the mixture, add two teaspoons of water and mix thoroughly.
2. Using a cellulose sponge, apply the green glaze to alternating squares. I usually stroke on a thin application of the glaze first. It will dry quickly and seal the tape edge. Then, when I rub and stroke the glaze on the floor, it will not bleed under the tape. Complete all green squares. Remove tape.
3. In another container, use the same recipe with light turquoise paint to mix the turquoise glaze.
4. Tape off the turquoise squares and use the same procedure as above to apply the glaze. (**photo 1**) Complete all squares and remove tape. Allow to dry thoroughly.

4 • Distressing & Softening

1. Using medium grit sandpaper, sand each square to give a worn look and reveal some of the wood.
2. *Optional:* Mix two parts gray white paint with two parts glazing medium. Add one ounce of water to each cup. Rub the glaze over the colors for a softer appearance. Let dry.

5 • Stamping

1. Trace the tulip design patterns onto tracing paper. Transfer them to stamping material using dressmaker's white carbon. Use a dressmaker's tracing wheel or a pen to transfer the lines. (**photo 2 and 3**)
2. Cut out the stamps. (**photo 4, 5, 6**)
3. Use a #12 artist brush to stroke paint colors on the flower shapes. Use white, sunflower yellow, lavender, and pink for the flowers. I used two colors per flower stamp for a more realistic effect. Stamp the tulips randomly on the floor, using photo as a guide for placement.
4. Stroke medium green and dark green on the leaf stamps, applying the paint in the direction of the leaf veins. (**photo 7**) Apply both colors at once onto the stamps, adding the dark color where the leaf might be shaded. Stamp the leaves. (**photo 8 & 9**)
5. Use a round brush to add stems and details with diluted paint. Let dry completely.

6 • Finishing

1. Lightly sand the stamped designs with fine (220 grit) sandpaper to create an aged, worn effect. Vacuum up dust.
2. Seal the surface with paste wax. Use a soft cloth and rub the wax on the painted wood, working it well into the surface. You will need to re-apply the wax for protection every year or two, depending on the traffic the floor receives. ✐

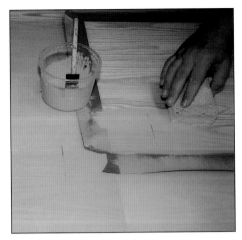

Photo 1. Color washing the turquoise squares.

Photo 2. Transferring the leaf design onto the foam material using a dressmaker's tracing wheel and white transfer paper.

Photo 3. Shows the transfer lines.

Photo 4. Cut out the design using a craft knife.

Photo 5. This photo shows the cut stamping pieces. Can hugger foam as shown at top right can also be used for stamping material.

Photo 6. Cutting detail lines into the stamp such as the vein line will result in giving the finished image dimension and detail.

Photo 7. Applying paint to the stamp.

Photo 8. Stamping the leaf image on the floor.

Photo 9. Lifting the stamp to reveal the image.

"La Richesse" Border
Stenciled Running Border on Dark Oak Floor

This border follows the perimeter of the room and accents the fireplace hearth. I used black paint to stencil the design, opting for a light application rather than solid coverage. The black color looks good with the deep mahogany brown of this oak floor. If you choose a lighter background stain, you may want to use a dark brown (or an even lighter color) for stenciling.

The pattern for this design is given after the projects chapter.

1 • Supplies

Stains, Paints & Finishes:
Oil-base wood stain: mahogany
Waterbase stencil paint: black
Finish of your choice

Tools & Equipment:
Chalk line
Stencil blank material
Stencil brush
Tracing paper and pencil
Cutting mat
Craft knife
Permanent fine tip marker

continued on page 46

Colors Used

mahogany oil-base wood stain

black stencil paint

continued from page 44

2 • Preparation

1. Hire a professional to sand the floor or do it yourself with a floor sander. Vacuum up dust.
2. Measure the position of the border and mark with a chalk line. (**photo 1**)
3. Plan the placement of the border. This is a running border with an exact continuous pattern, so it's important to determine where the pattern will be placed so the corners will look good. An easy way to do this is to trace the pattern from the book and make multiple photocopies of the design. Place the copies along the floor on one side of the room at a time. Adjust the design as needed. Study the corners in the photos to see adjustments that were made in the design to fit this room.

3 • Making the Stencil

1. Trace the stencil patterns on stencil blank material with a fine tip marker.
2. Place the stencil blank material on the glass. Using a craft knife, carefully cut out the designs.

4 • Stenciling

1. Position the stencil design, centering it on the chalk mark.
2. Stencil the design with black stencil paint, applying the paint with soft, smooth circular strokes and allowing the wood grain to show through. (**photo 2, photo 3**) Let dry completely.

5 • Finishing

Seal the floor. ✆

Photo 1. Marking the placement with a chalk line.

Photo 2. Stenciling the first overlay.

Photo 3. The design after the first overlay; note how the design was adjusted at the corner.

Country French Stripes & Spatters
Painted Wood Kitchen Floor

This bright floor uses the floor boards as guides for creating stripes of various widths. Harmonizing paint colors are blended on the stripes for a streaky effect. All the floor colors are used for spattering. This design would work well on a concrete floor, an asphalt floor, or a plywood floor.

1 • Supplies

Paints & Finishes:

Acrylic paint: ultramarine blue, brilliant blue, tangerine, sunflower yellow, light yellow

Finish of your choice

Tools & Equipment:

Masking tape
Medium grade sandpaper
Bristle paint brushes
Screwdriver

Paint Colors Used

ultramarine blue

brilliant blue

tangerine

sunflower yellow

light yellow

2 • Preparation

Prepare the wood floor for painting.

3 • Painting

1. Use masking tape to section off the stripes—some are the width of one board only; others cover two or three boards.
2. The blue stripes were created by dipping the brush first in one of the blue colors, then in the other. When the paint is applied, there is a subtle streaked effect.
3. The yellow stripes were painted the same way. Some boards had three colors; others had two. Remove tape and allow the stripes to dry before painting adjacent boards. When all painting is complete, allow to dry thoroughly.

4 • Sanding

Sand the surface, using medium grade sandpaper. I sanded one color first, then the other. Vacuum up all sanding dust before sanding the other color. **Always** change sandpaper when moving to another color—the paint on the sandpaper will mar the surface if you don't.

5 • Spattering

1. Working one color at a time, dilute the paint with water, using one part water to two parts paint. Dip the bristles of a brush into the diluted paint, hold the brush over the surface, and tap the shank of the brush with a screwdriver. This causes the bristles to spatter paint on the floor. Work from one side of the floor to the other. Allow to dry completely.
2. Dilute the other colors, one at a time, and repeat the procedure. Allow to dry thoroughly after each spattering.

6 • Finishing

Seal the surface. ✪

Granite & Wood
Wood Floor with Faux Stone Inlays

The wooden floor in this kitchen is given a contemporary look with a silver metallic-painted grid (faux stainless steel) and granite-look sponged inlays and a border. The border camouflages the damaged edges of the floor where linoleum had once been tacked down with nails that rusted and discolored the wood.

1 • Supplies

Stains, Paints & Finishes:

Waterbase wood stain: white (also called "pickling stain" or "whitewash")

Acrylic craft paint: metallic silver, black, medium gray, white, tan

Finish of your choice

Tools & Equipment:

Brushes, sponges, and rags for staining

Chalk line

Measuring tape and ruler

Masking tape - 1/4", 1", 2"

Sponge brush

Stencil blank material

Tracing paper and pencil

Cutting mat

Craft knife

Permanent fine tip marker

Stencil brush

Natural sponge

Paint Colors Used

black

white

medium gray

tan

metallic silver

2 • Preparation

1. Prepare the wood floor.
2. Stain the entire floor white. (**photo 2 on page 52**) Allow to dry.

3 • Creating the Grid

1. Measure and mark the grid, using sizes appropriate for your room. Here, the border is 4" wide, the small granite inlay squares are 3", and the grid divides the wood floor into rectangles 18" x 15". Use a chalk line to mark the rectangles.
2. Position 1/4" masking tape along the lines to mark the grid. Place 1" masking tape along each side of the 1/4" tape. Carefully remove the 1/4" tape and press the wider masking tape securely to the floor. Work first in one direction, then the other.
3. Paint the 1/4" wide grid with silver metallic paint. (This will resemble stainless steel.) (**photo 3**) Remove tape and allow to dry. Tape and paint the lines in the other direction.

4 • Creating the Border

Practice sponging the faux granite on a piece of wood or posterboard before working on your floor.

1. Mask off around the border and place masking tape on the floor moldings to protect them. In doorways, I positioned masking tape along the center line of each doorway and painted the granite faux finish up to the center of the doorway.
2. Paint the border black. Allow to dry.
3. Sponge with medium gray. Do not sponge so heavy as to cover the black background color.
4. Sponge *lightly* with tan.
5. Sponge with white.

5 • Creating the Inlays

1. Cut a 3" (or the size you wish) square from a piece of stencil blank material.
2. Tape stencil in place where grid lines intersect. Sand inside the square with sandpaper. (**photo 4**) Vacuum up sanding dust.
3. Stencil the square with black. Remove stencil. Let dry. Repeat until complete. (**photo 6**)
4. Tape off around each square.
5. Sponge the squares with medium gray. (**photo 7**)
6. Sponge the squares *lightly* with tan.
7. Sponge the squares with white. Remove tape. (**photo 8**)

6 • Finishing

Seal the surface. (**photo 9**) ☙

Photo 1. Before decorating. The vinyl flooring and glue residue had to be removed before the floor was sanded.

Photo 2. Applying white stain to the floor with a sponge.

Photo 3. Painting the grid with silver metallic paint, using a sponge brush.

Photo 4. Sanding the inlay squares.

Photo 5. Painting the inlay squares with a stencil brush.

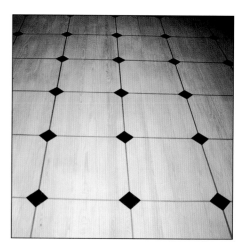

Photo 6. The inlays painted, before sponging.

Photo 7. Sponging the inlay squares.

Photo 8. The inlay, with sponging complete.

Photo 9. Applying the finish. A mop with a lamb's wool cover makes this an easy job.

Compass Points
Stained & Stenciled Floor

This modern design, accomplished with three stain colors on a natural wooden floor, is a perfect base for this home office. The wood tones help warm up what could be a cold, hard room. The stenciled medallions add the finishing touch.

The pattern for this floor is at the end of the projects chapter.

1 • Supplies

Stains & Finishes:
Waterbase wood stains: mahogany, pecan, walnut
Finish of your choice

Tools & Equipment:
Masking tape, 3"
Chalk line

Stain Colors Used

mahogany

pecan

walnut

2 • Preparation

1. Prepare the wood floor. I did not stain the floor, I used the natural wood color after sanding as the background.
2. Measure your floor space and divide into squares of equal size. Mark and position 3" masking tape to make a grid. (**photo 1 on page 56**) You'll leave this tape in place until you stain the points.

3 • Stain the Triangles

1. To create the triangles, use a chalk line to mark a diagonal line from corner to corner. (**photo 1**)
2. Use tape to mask off the two opposite triangle shapes. Stain these triangles with pecan stain. (**photo 2**) Remove tape. Use tape on the remaining two triangles and stain them walnut.
3. Remove the tape from around the triangle and let dry.
4. Continue to mask off and stain until all the triangles are complete.

4 • Stain the Squares

1. Use a straight edge and a craft knife to very carefully cut away the tape where it intersects to form the squares. (**photo 3**)
2. Stain the squares with walnut stain. (**photo 4**) Let dry.
3. Remove all the tape. (**photo 5**)

5 • Stain the Points

1. Trace the pattern provided on stencil blank material with a fine tip marker.
2. Using a craft knife, carefully cut out the design.
3. Position the stencil around the squares. Stencil the points with mahogany. (**photo 6**) Allow to dry.

6 • Finishing

Seal the surface. ℘

1. Snapping the chalk lines for the triangles.

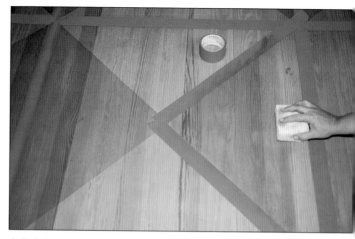

2. Staining two opposite triangles with walnut, and the two remaining triangles with pecan..

3. Peeling away the tape to reveal a square.

4. Staining a square with walnut.

5. Remove the grid tape.

6. Staining the points with mahogany.

◆

Faux Inlay
Stained Floor Using Colored Stain

In this living room, a floral design was color stained on the floor to create the look of an inlaid border. The heart pine floor was not in perfect condition, so before the border was applied, the floor was repaired and sanded, and a wood-tone stain was applied to the entire floor to even out the color. The design is hand painted with an artist's brush. I applied a tung oil finish to enhance the aged appearance.

Patterns for this project are on page 61.

1 • Supplies

Stains, Paints & Finishes:
Oil-base or waterbase wood-tone stain
Waterbase paint: dark brown, light brown, dark green, deep red
Glazing medium
Tung oil finish

Tools & Equipment:
Artist's paint brush, #8 round
Tracing paper
Transfer paper
Tape measure or ruler
Pencil
Chalk line and red chalk powder
Disposable cups
Low tack masking tape
Cloth rags
Rubber gloves
Optional: Sponge mop

Paint Colors Used

dark green

deep red

light brown

dark brown

continued on page 60

Photo 1. Before

Photo 2. Staining the floor with a sponge mop.

Photo 3. Masking off the border.

Photo 4. Transferring the pattern.

Photo 5. Staining the border with dark brown.

Photo 6. Painting the design with a brush.

Continued from page 58

2 • Preparation

1. Hire a professional to sand the floor or do it yourself with a floor sander. Vacuum up the dust.
2. Apply the wood-tone stain to the floor. I used a sponge mop to apply the stain (**photo 2**). Allow to dry.
3. Measure 3" from the wall and tape off a stripe 1-1/2" wide. Tape off a second 1-1/2" stripe 21" from the wall (**photo 3**). Use a pencil to mark the measurements and a chalk line to mark the lines for taping.
4. Enlarge the patterns for this project on a copy machine.

5. Trace the photocopied design on tracing paper. Tape the tracing paper to the floor. Slip transfer paper between the floor and the tracing paper. Go over the lines to transfer the pattern to the floor (**photo 4**).

3 • Color Staining

1. Using a separate cup for each paint color, dilute each paint color with water or glazing medium to make a transparent stain, using 3 parts paint to 1 part glazing medium. Test the mixtures on a scrap of wood before applying the color stain to your floor.

2. Stain the stripes with dark brown (**photo 5**). *NOTE: Some of the floral design overlaps the taped stripes. When you stain the stripes, be sure to leave a 1/4" space between the design and the stripe.* Allow to dry. Remove tape.
3. Paint the floral design with light brown, dark green, and deep red. Use photos as guides for color placement. Be sure to stroke the color enough to get a transparent look (**photo 6**). Allow to dry.

4 • Finishing

Apply a tung oil finish to the floor, following manufacturer's instructions. ☙

After. Use this photo as a pattern. Trace from the book and enlarge to desired size.

Trace pattern from book and enlarge to desired size for corners.

Centerpiece Medallion
Stenciled Medallion on a Stained Floor

The center medallion stenciled on this floor accents the placement of the table and chairs. The black stencil paint creates a subtle design on the dark floor and allows the wood grain to show.

The pattern for this floor is included on page 65.

1 • Supplies

Stains, Paints & Finishes:
Oil-base stain: mahogany
Waterbase stencil paint: black
Finish of your choice

Tools & Equipment:
Chalk line
Stencil blank material
Stencil brush
Tracing paper and pencil
Cutting mat
Craft knife
Permanent fine tip marker

2 • Preparation

1. Hire a professional to sand the floor or do it yourself with a floor sander. Vacuum up dust.
2. Stain the floor with mahogany stain. (**Photo 2 on page 64**)
3. Locate the center of the room by marking with a chalk line.
4. Plan the placement of the medallion. The one shown here is 48" in diameter. Enlarge the design, if needed, to fit your room or your table.

3 • Making the Stencil

1. Trace the pattern on stencil blank material with a fine tip marker. To do this design, I cut one "spoke" of the design and the circle on one of the stencil overlays. For the other stencil overlay, I cut the outside curve for the area between the "spokes."
2. Place the stencil blank material on the glass. Using a craft knife, carefully cut out the design.

4 • Stenciling

1. Position the stencil design, centering it on the chalk mark.
2. Stencil the design with black stencil paint, applying the paint with soft, smooth circular strokes and allowing the wood grain to show through. Stencil heavier around the edges of each spoke. I used the "spoke" overlay over and over, stenciling the design around the circle until the medallion was complete. Be sure to measure and place it evenly around. Let dry completely.
3. Place tape over the edges of each side of the spokes.
4. Position the stencil overlay curve. Stencil and shade around the outside edges of each spoke and inside the curve.
5. Allow to dry.

5 • Finishing

Seal the floor. ☙

Photo 1: Here is the room shown before decorating. Most of the sanding has been done to remove the old stain— only a small corner area is yet to be sanded.

Photo 2: Apply mahogany stain with a cloth.

Pattern for
Centerpiece Medallion

Actual size is 48" diameter.
Enlarge to desired size.

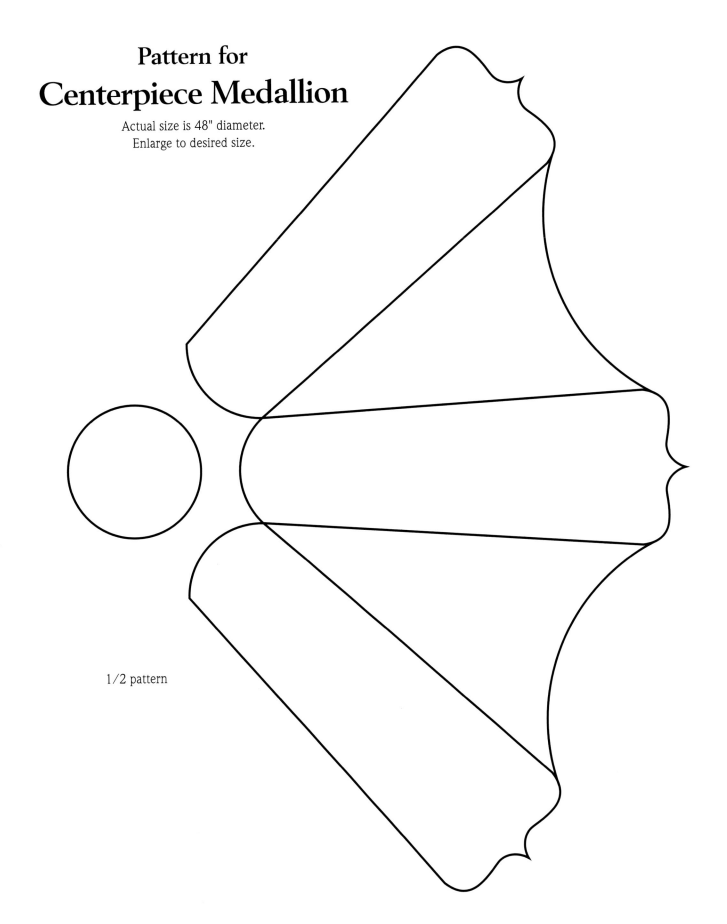

1/2 pattern

Mountain Laurel
Stenciled Deck Floor on a Screened Porch

The deck flooring of the screen porch of this mountain cabin is randomly decorated with clusters of mountain laurel. The flooring was left its natural color. Waterbase exterior paint was used for stenciling.

The pattern for this floor is included on page 68.

1 • Supplies

Paints

Acrylic outdoor craft paint: fairway green, olive, green mist, cumin, white

Tools & Equipment:

Stencil blank material
Stencil brushes
Tracing paper and pencil
Cutting mat
Craft knife
Permanent fine tip marker

Paint Colors Used

fairway green

olive

green mist

cumin brown

white

2 • Preparation

1. Wash the wood decking and let dry completely.
2. Plan the placement of the random designs. Enlarge the designs, if needed, to fit.

3 • Making the Stencil

1. Trace the pattern on stencil blank material with a fine tip marker. Because I placed and stenciled the leaves and flowers randomly along the stem to make each branch look different, I cut this stencil slightly different. One stencil overlay was cut for the large leaf; one stencil overlay was cut for the small leaf; and one stencil overlay was cut for the stem.
2. Using a craft knife, carefully cut out the designs.

4 • Stenciling

1. Position the stem design and tape in place. Stencil the design with fairway green and shade in places with cumin brown.
2. Stencil the leaves along the stem, stenciling some large and some small. Stencil them in a variety of places along the stem so each stem looks different. When stenciling the leaves, I used the stencil bristle brush to stroke the three green colors randomly in the direction of the leaf veins. I used the cumin brown to shade some of the leaves.
3. The flowers were stenciled randomly in clumps along the stem. The paint was applied with circular strokes and also pounced so the paint is uneven. A dot of green was added to the center of some of the flowers using a small brush.
4. Let dry. No sealer is required. ✍

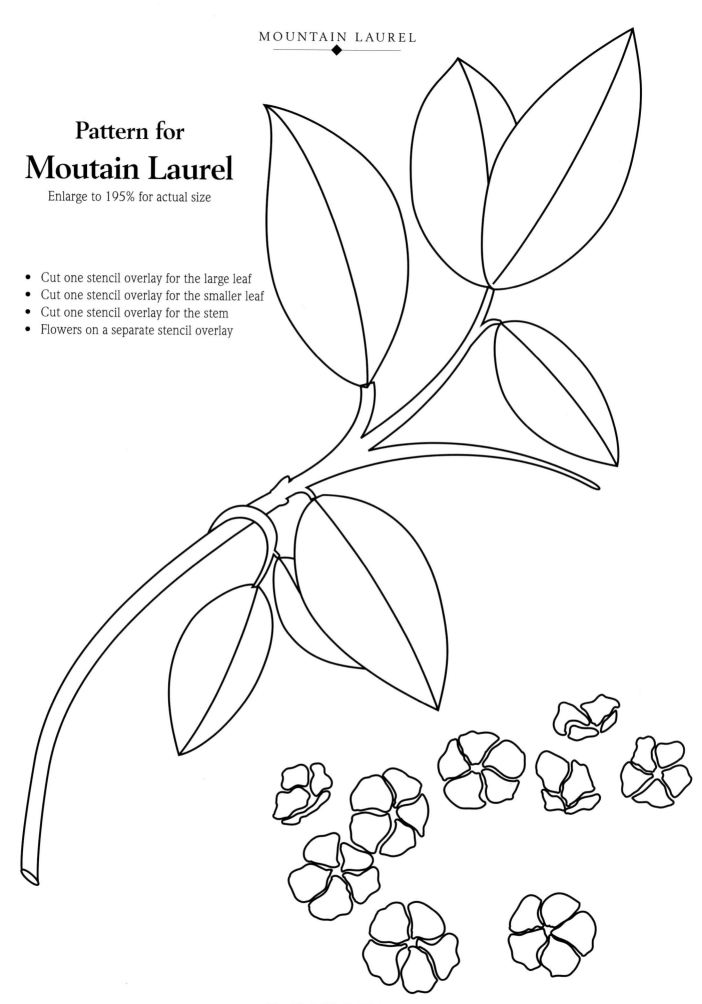

Pattern for
Moutain Laurel
Enlarge to 195% for actual size

- Cut one stencil overlay for the large leaf
- Cut one stencil overlay for the smaller leaf
- Cut one stencil overlay for the stem
- Flowers on a separate stencil overlay

Leaf Strewn Deck
Stenciled & Woodburned Leaves on an Exterior Deck

Green leaves are stenciled randomly on this outdoor deck and accented with woodburned details. For added realism, use actual leaves from your yard, your garden, or the woods to make the stencil patterns. I made my leaves oversized to make them more of a statement. You can enlarge leaves on a copier to get the size needed. We have included some patterns of the leaves we used. Use them if you can't find leaves you like in your own yard.

Patterns for this project are on page 72.

1 • Supplies

Paints & Stains:

Acrylic outdoor craft paint: fairway green, olive, dark brown

Tools & Equipment:

Real leaves (*Option:* leaf patterns are supplied.)
Woodburning tool
Stencil blank material
Stencil brushes
Tracing paper and pencil
Cutting mat
Craft knife
Permanent fine tip marker

Paint Colors Used

fairway green

olive

dark brown

2 • Preparation

1. Clean the wood decking. Allow to dry thoroughly. This deck is a natural weathered color, with no stain or sealer used on it.
2. Decide where you want to stencil the leaves and place little pencil marks.
3. Learn about the woodburning tool. Read the instructions that come with the woodburning tool. The tool comes with different tips—choose the appropriate one for the various design elements. Practice with the tool on scrap lumber; especially practice woodburning on the cross grain of the wood. When working across the grain, it's more difficult to achieve thin, straight lines, but you can do it if you work slowly and carefully. Practice helps!

3 • Making the Stencil

1. Trace the leaves on stencil blank material with a fine tip marker. Use as many different designs as you like.
2. Place the stencil blank material on the glass. Using a craft knife, carefully cut out the designs.

4 • Stenciling

1. Position the stencil design and tape in place.
2. Stencil the leaves. I used the two green colors to stencil randomly in the stencil opening. Use a light hand—you don't want solid coverage. Add touches of dark brown for shading. Let dry.

5 • Woodburning

Use the woodburning tool to outline and shadow some leaves and to create the veins and stems. Refer to the real leaves or the patterns you used to make the stencils. You can trace the veins on tracing paper, transfer them to the stenciled leaves, and woodburn them or simply "draw" them freehand with the woodburning tool. (I did not always reproduce every vein and stem—on some leaves, it would have looked too busy.) No sealer is necessary. ☙

Patterns for
Leaf Strewn Deck
Enlarge to 235% for actual size

Sisal Beauty
Stenciling on Sisal

This stenciled sisal rug anchors the seating area in this garden room. The blue border and trellis grid were marked with masking tape and painted. The stenciled flowers and vines—a precise, repeated design—seem to grow around the band, giving a dimensional look to the flat surface.

The pattern for this floor is at the end of the projects chapter.

1 • Supplies

Paints & Finishes:
Acrylic craft paint: French blue, blue ribbon, shamrock, rose pink, rose garden, white
Spray rug protector

Project Surface:
Woven sisal rug, 9' x 12'

Tools & Equipment:
Masking tape
Stencil brushes
Stencil blank material
Tracing paper and pencil
Cutting mat
Craft knife
Permanent fine tip marker
Optional: Airbrush, for painting the border

Paint Colors Used

French blue

blue ribbon

shamrock

rose pink

rose garden

white

2 • Preparation

1. Unroll the sisal rug. Weight the corners with heavy objects (large books are useful) and let it relax until it is perfectly flat.
2. Secure it to your working surface so it will not curl up or move while you work.
3. Plan the placement of the design on your rug. Enlarge the stencil design as needed.

3 • Painting the Border

1. Measure in 8" from the edge of the rug. Mask off a band 10" wide and press the tape down firmly.
2. Paint the band with French blue. Use either a stencil bristle brush, stippling the color into the rough fibers and making sure not to use so much paint on the brush so that the paint bleeds under the masking tape, or use an airbrush.

4 • Making the Stencil

1. Trace the pattern on stencil blank material with a fine tip marker. This design is fairly complicated. Cut one stencil overlay for the green leaves. Cut one stencil overlay for the blue vine. Cut one stencil overlay for the light pink flowers — this will include some small flowers, the complete outside outline of the large flower, and the complete outline of the medium flower. Cut one overlay for some small white flowers—include on this overlay the shaded areas of the medium and large flowers.
2. Place the stencil blank material on the glass. Using a craft knife, carefully cut out the designs on the four overlays.

5 • Stenciling

1. Stencil the vine using blue ribbon (dark blue). Stencil the vine around the rug, weaving it through and around the light blue band.
2. Stencil the leaves using the shamrock green color.
3. Stencil the light pink overlay. Use rose pink to stencil in some of the small flowers, the entire medium and large flowers. While the stencil is in place, use the stencil brush to shade the large flower here and there with the dark pink color (rose garden).
4. Position the next overlay and stencil the small white flowers and the area on the large flower. Load another brush with dark pink (rose garden) and stencil the shaded area of the medium flower as well as shading some places on top of the white area of the large flower.

6 • Creating the Grid

The diamond grid is painted with alternating stripes of French blue and blue ribbon. (See photo as a guide for color placement.) Use masking tape to section off the lines. Stipple each color, using a stencil brush.

7 • Finishing

Seal the surface with a spray rug protector. Apply several coats. You may want to apply a non-skid brush on backing to the rug if it does not already have one. Or you can use a foam rug pad under the rug. ✆

Country Comfort
Stenciling on Sisal

This is a very easy stenciling project, yet it can add a huge amount of country warmth and style to an area of your home. The sisal area rug is simply decorated with a wide border and a stenciled vine. The primary color scheme is versatile and would work with rustic country furnishings as well as a contemporary décor.

1 • Supplies

Paints & Finishes:
Acrylic craft paints: navy blue, red
Spray rug protector

Project Surface:
Woven off-white sisal rug, 6' x 9'

Tools & Equipment:
Pre-cut stencil of a vine
Masking tape
Stencil brushes

2 • Preparation

1. Unroll the sisal rug. Weight the corners with heavy objects (large books are useful) and let it relax until it is perfectly flat.
2. Secure it to your working surface so it will not curl up or move while you work.
3. Plan the placement of the stenciled design on your rug. Enlarge or adjust pattern as needed.

3 • Painting the Border

1. Measure in 2-1/2" from the edges and mask off a band 2" wide. Press the tape down firmly.
2. Paint the band with red, using either a stencil bristle brush, stippling the color into the rough fibers and being sure not to use so much paint on the brush so that it bleeds under the masking tape, or use an airbrush. Let dry.

4 • Stenciling

Stencil the vine pattern on the rug using navy blue. Stipple with a brush or use an airbrush.

5 • Finishing

Seal the rug with a spray rug protector. Apply several coats. ☙

Paint Colors Used

navy blue

red

Magnolia Elegance
Stenciling on Canvas Floorcloth

By Donna O'Rourke Mabrey

This canvas floorcloth, with an outer border of metallic gold and stenciled magnolia blossoms, creates an elegant focal point for this hallway.

The pattern for this floorcloth is at the end of the projects chapter.

1 • Supplies

Paints:

Indoor/outdoor acrylic paint: vanilla, 1 qt

Acrylic craft paint: linen, white, almond, warm brown, teal, teal green, avocado, maroon, metallic gold

Neutral waterbase glazing medium

Polyacrylic semi-gloss protective finish

Polyacrylic satin protective finish

Project Surface:

Floorcloth canvas, 6' x 9'

Tools & Equipment:

Sea sponge

Masking tape (low-tack type)

Measuring tape

Yardstick

Paint brushes

Stencil brushes

Stencil blank material

Tracing paper and pencil

Cutting mat

Craft knife

Permanent fine tip marker

Stenciling Option: Purchase pre-cut magnolia, scroll, and rope border stencils.

Paint Colors Used

vanilla

linen

white

almond

warm brown

teal

teal green

avocado

maroon

metallic gold

2 • Preparation

1. Cut the canvas to size and hem. See "Basic Steps for Making a Canvas Floorcloth." Allow glue to dry thoroughly.
2. Paint entire front of canvas with two coats vanilla. Let dry between coats. Let second coat dry completely.

3 • Sponging & Borders

1. Mix equal amounts linen paint and neutral glazing medium. Dampen sea sponge. Sponge glaze mixture over the surface to create a mottled look. Let dry.
2. Mask off a 4" border on the floorcloth. Paint the border with gold metallic. Remove tape. Let dry.
3. Measure in 1" and mask off a band 1/2" wide. Paint with teal. Remove tape. Let dry.

4 • Making the Stencils *If you're using pre-cut stencils, move on to part 5.*

1. Trace the patterns from the book and enlarge on a copy machine to size noted or size desired.
2. Trace the enlarged patterns on stencil blank material with a fine tip marker. You will need the following stencil overlays:
   - One stencil overlay for **scroll border**
   - One stencil overlay for **magnolia petals**
   - One stencil overlay for **leaves and stems**
   - One stencil overlay for the **rope border**

   Note: The scroll design is used as a mirror image on either side of the magnolias. You can cut two different scroll stencils or simply use both sides of the stencil to get a mirror image. If you choose to use both sides, clean the stencil thoroughly before turning it over on the canvas.
3. Place the stencil blank material on the glass. Using a craft knife, carefully cut out the designs.

5 • Stenciling

1. Position the rope border stencil between the band and the border. Stencil with maroon.
2. Stencil the leaves overlay using avocado stenciled lightly. With stencil in place, shade the leaves and stems with teal green. with stencil still in place, shade leaves and stems darker in areas with warm brown.
3. Stencil the magnolias with white. With stencil in place, shade petals with almond and then with linen.
4. Stencil the scrolls with teal. Let dry completely.

6 • Finishing

1. Apply 4 coats polyacrylic gloss allowing to dry between coats.
2. Finish with a coat of polyacrylic satin or matte. ☙

Harvested Fruits
Stenciling on Sisal Rug

By Donna O'Rourke Mabrey

We've shown this sisal rug as a setting for an outdoor picnic, but it would be equally at home in a dining room, kitchen or breakfast room, an entry hall, or porch. The fruit border design could work on a rug of any size.

The pattern for this floorcloth is at the end of the projects chapter.

1 • Supplies

Paints & Finishes:
Acrylic craft paint: honeycomb, seafoam, emerald, blueberry, maroon
Spray rug protector

Project Surface:
Sisal rug, 6' x 9'

Tools & Equipment:
Stencil brushes
Stencil blank material
Tracing paper and pencil
Cutting mat
Craft knife
Permanent fine tip marker
Optional: Airbrush, for painting the border

Paint Colors Used

honeycomb

seafoam

emerald

blueberry

maroon

2 • Preparation

1. Unroll the sisal rug. Weight the corners with heavy objects (large books are useful) and let it relax until it is perfectly flat.
2. Secure it to your working surface so it will not curl up or move while you work.
3. Plan the placement of the stenciled design on your rug. Enlarge or adjust patterns as needed.

3 • Making the Stencils

1. Trace the patterns on stencil blank material with a fine tip marker. You will need stencil overlays for the following:
   One stencil overlay for the twining ribbon border
   One stencil overlay for leaves and stems. Cut the complete leaf outline, disregarding the shaded areas on the leaves—these will be cut as a separate overlay to be stencils on top of the light leaf color.
   One overlay for the shaded sections on leaves
   One overlay for the plums and the grapes (these can be put on the same overlay—but can be cut as separate overlays if you prefer)
   One overlay for the pears
2. Place the stencil blank material on the glass. Using a craft knife, carefully cut out the design.

4 • Stenciling

1. Stencil the leaves using seafoam green color. With this stencil in place, shade around the edges of the leaves with emerald. Stencil the overlay with the shaded section of the leaves using emerald.
2. Stencil the grapes using blueberry.
3. Stencil the plums using maroon.
4. Stencil the pears using Honeycomb. Stencil the edges a little heavier to create a shaded look.
5. Stencil the border with emerald paint to connect the fruit designs. *Option:* Use an airbrush for this. Let dry completely.

5 • Finishing

Seal the rug with a spray rug protector. Apply several coats. ✍

Patio Medallions
Stenciled Concrete Stepping Stones

These stenciled concrete stepping stones can be arranged indoors or out to create an interesting patterned floor. Outdoors, gravel can be used between them. I saw a similar idea done in a bathroom using smooth pebbles between the blocks. It had a very Zen-like quality.

The pattern for these stepping stones is at the end of the projects chapter.

1 • Supplies

Paints:

Acrylic craft paint suitable for outdoor use: fairway green, midnight, oxblood

Project Surface:

Square cement or concrete stepping stones, 18" square

Tools & Equipment:

Stencil brushes
Stencil blank material
Tracing paper and pencil
Cutting mat
Craft knife
Permanent fine tip marker

2 • Making the Stencil

1. Trace the Medallion Floorcloth patterns from page 125 (center medallion, corners, and edge shapes) and enlarge on a copy machine as needed.
2. Trace the enlarged patterns on stencil blank material with a fine tip marker. It is easiest to cut this stencil in one overlay rather than having a separate overlay for each color. Because the cutout areas are small, you can simply tape over areas on the stencil with masking tape when stenciling adjacent areas of different colors.
3. Place the stencil blank material on the glass. Using a craft knife, carefully cut out the designs.

3 • Stenciling

You can have fun with this stencil by positioning the stencil elements differently for each cement square. Use the three colors listed interchangeably. Follow the example in our photograph. No sealing or other finishing is necessary. ✆

Medallion Floorcloth
Stenciling on Back of Vinyl Flooring Remnant

This octagonal floorcloth is stenciling on the back of a piece of vinyl flooring. You can purchase vinyl flooring inexpensively at building supply stores and home improvement centers. It can be cut to any shape, is easy to clean, and won't slip on your floor.

Patterns for this project are at the end of the Projects chapter.

1 • Supplies

Paints & Finishes:

Indoor/outdoor paint: beige or khaki, 1 qt.
Acrylic craft paint: black, cumin, fairway green, pediment gray, oxblood
Polyacrylic semi-gloss protective finish
Polyacrylic satin protective finish

Project Surface:

Vinyl flooring, 5-1/2 ft. square

Tools & Equipment:

5 stencil brushes, 3/8"
Paint roller and pan
Foam brush, 2"
Sea sponge or sea sponging mitt
Masking tape, (width of grout)
Low-tack masking tape
Sandpaper, 100 or 150 grit
Cloth rag
Ruler
Disposable plate
Stencil blank material
Tracing paper and pencil
Cutting mat
Craft knife
Permanent fine tip marker

Paint Colors Used

black

cumin

pediment gray

oxblood

fairway green

2 • Preparation

1. Measure, mark, and cut the vinyl flooring. Cut the vinyl on a garage floor or area that will not get damaged when the blade of a craft knife cuts through it. Create the octagonal shape this way: Measure 18" from both sides of a corner of the square piece of flooring. Place the ruler across the corner, mark, and cut off this corner. Repeat on all corners.
2. Sand the back of the octagon to allow the paint to adhere and bond to the surface. Remove sanding dust with a cloth rag.
3. Paint this surface with one to three coats of beige or khaki, using the paint roller. Allow the paint to cure for 24 to 48 hours.

3 • Sponging

Moisten the sea sponge or face of the sponging mitt. Pour some cumin paint on a disposable plate and dip the sponge in the paint to load. Sponge the khaki surface of the octagon with cumin, leaving some areas darker than others. Let dry thoroughly.

4 • Making the Stencil

1. Trace the patterns from this book (center medallion, corners, and edge shapes) and enlarge on a copy machine as needed.
2. Trace the enlarged patterns on stencil blank material with a fine tip marker. It is easiest to cut this stencil in one overlay rather than having a separate overlay for each color. Because the cutout areas are small, you can simply tape over areas on the stencil with masking tape when stenciling adjacent areas of different colors.
3. Place the stencil blank material on the glass. Using a craft knife, carefully cut out the designs.

5 • Stenciling

1. Stencil the border of the floorcloth, starting in the center of each side edge of the cloth and stenciling outward in both directions. Use the photo as a guide for color placement.
2. Measure and mark the center of the floorcloth. Position the medallion stencil in the center and stencil the design with the colors shown in the photo.
3. Position the stencils for the black edge shapes all along the edge of floorcloth. Stencil these areas with black. Let dry.

6 • Creating Connecting Rays

1. Use 1/4" tape to create the ray pattern. Position strips of tape to connect the center medallion with the border. Place a strip of masking tape along each side of the 1/4" tape. Secure the inner edges by pressing with your fingers, then remove the 1/4" tape. Place small pieces of tape on each end of each ray.
2. Paint the rays with black. Remove tape.

7 • Finishing

1. Apply 4 coats polyacrylic gloss allowing to dry between coats.
2. Finish with a coat of polyacrylic satin or matte. ✍

Luscious Lemons
Painted Canvas Floorcloth

By Patty Cox

Bright yellow lemons adorn the center of this painted floorcloth. The design is easy to paint with a resist technique, cutting the shapes from freezer paper and ironing them in place before painting the background.

Patterns for this project are at the end of the Projects chapter.

1 • Supplies

Paints & Finishes:
Acrylic craft paint: violet pansy, purple lilac, sunny yellow, black, lipstick red, apricot, periwinkle blue, green clover
Neutral waterbase glazing medium
Polyacrylic semi-gloss protective finish
Polyacrylic satin protective finish

Project Surface:
White floorcloth canvas, 38" x 31" (finished size: 36" x 29")

Tools & Equipment:
Masking tape
1" sponge brush
#6 liner brush
Sponge
Freezer paper
Wash-away fabric marker
Iron

2 • Preparation

1. Prepare canvas and turn edges. See "Basic Steps for Making a Canvas Floorcloth." If your canvas is not white, you will need to paint it with white gesso.
2. Measure and mark canvas according to Fig. 1 on page 88 using a wash-away fabric marker.

continued on page 87

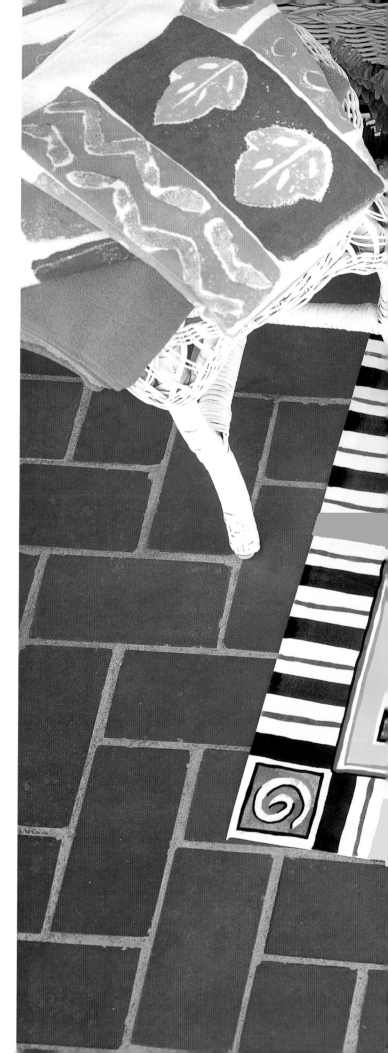

continued from page 86

3. Mask 1-1/2" inner border with two rows of masking tape.

4. Trace 5 lemons, 7 leaves, 7 spirals and 6 small daisy patterns from book onto freezer paper. Freezer paper is see-through enough that you can place the paper over the patterns and trace. Cutout the shapes from freezer paper. Arrange on floorcloth center with wax side down, using photo as a guide. Arrange a spiral in each corner. Press freezer paper in position with an iron.

3 • Painting

1. Sponge center area over freezer paper shapes with violet pansy, purple lilac, and periwinkle.

2. Paint black wash stripes around outer border with a 1" sponge brush.

3. Using a #6 liner brush, paint periwinkle blue wash stripes between black stripes on outer border. Allow paint to dry. Remove masking tape.

4. Paint an orange (red + apricot) square in each corner over spiral freezer paper. Lightly sponge apricot over each square.

5. Remove all freezer paper shapes.

6. Paint lemons with a yellow wash, using a sponge brush and leaving an unpainted white highlight on each. Shade side opposite highlight with an apricot wash.

7. Paint a yellow wash circle in the center of each daisy. Shade with apricot.

8. Paint a 1" apricot wash border around center sponged area, leaving area around leaves unpainted.

9. Paint leaves a green + yellow wash with sponge brush, leaving an unpainted white highlight on each leaf. Shade with a wash stroke of green.

10. Mix red + yellow, making orange. Paint an orange line around apricot border, using a liner brush. Leave a narrow white space and paint a red line just outside of the orange line.

11. Dip a pencil eraser in red paint. Stamp red dots between lemon designs on purple background.

12. Outline lemons, leaves, daisies and borders black using a liner brush. Paint a black drop shadow on each spiral. Let dry.

4 • Finishing

1. Apply 4 coats polyacrylic gloss allowing to dry between coats.

2. Finish with a coat of polyacrylic satin. ✍

Fig. 1

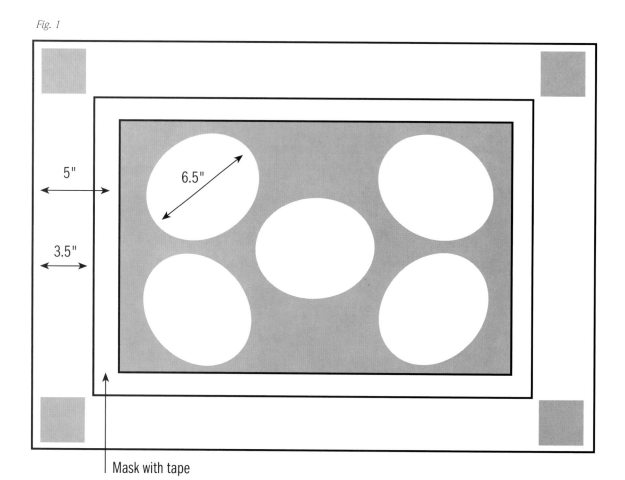

5"

6.5"

3.5"

Mask with tape

Patterns for
Luscious Lemons

Comparing Pears & Oranges
Painted Canvas Floorcloth

By Patty Cox

Bright washes of color are used to paint these pears and oranges. Some areas are left unpainted, so the white of the canvas shows through, creating highlights.

Patterns for this project are on page 93.

1 • Supplies

Paints & Finishes:
Acrylic craft paint - yellow, orange, light green, crimson, magenta, ultramarine, purple, black
Neutral waterbase glazing medium
Polyacrylic semi-gloss protective finish
Polyacrylic satin protective finish

Project Surface:
White floorcloth canvas, 38" x 30" (finished size: 36" x 28")

Tools & Equipment:
1" and 2" sponge brushes
#6 liner brush
wash-away fabric marker

continued on page 92

continued from page 90

2 • Preparation

1. Prepare canvas and turn edges. See "Basic Steps for Making a Canvas Floorcloth."
2. Measure and mark canvas according to Fig. 1, using wash-away fabric marker. Patterns are provided for transferring to canvas.

3 • Painting

ALL painting on this floorcloth is done with a wash (paint color + water).

1. Paint 2" border stripes with a mix of crimson + magenta using a 2" sponge brush.
2. Paint pears with a mix of yellow + a little orange, leaving white highlights unpainted. Add more orange to wash to shade pears.
3. Use the same orange wash to paint oranges, leaving highlights unpainted. Add more orange to wash mix. Shade oranges with this darker mixture.

4. Paint triangles with the darker orange mix.
5. Paint leaves with a mix of light green + a little yellow, leaving white highlights unpainted. Shade each with a darker green wash. Allow washes to dry.
6. Paint outer border background with purple, leaving white space around each orange.
7. Paint center background with ultramarine, leaving white space around each motif. Let dry.
8. Outline each motif with black, using a liner brush.
9. Mix only a small amount of water with black paint. Paint a dimple in each orange. Paint center vein on each leaf. Paint a stem on each pear. Let dry.

4 • Finishing

1. Apply 4 coats polyacrylic gloss, allowing to dry between coats.
2. Finish with a coat of polyacrylic satin or matte. ✥

Fig. 1

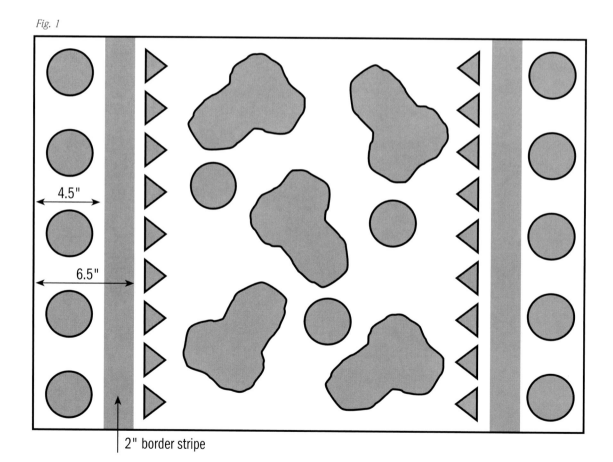

4.5"

6.5"

2" border stripe

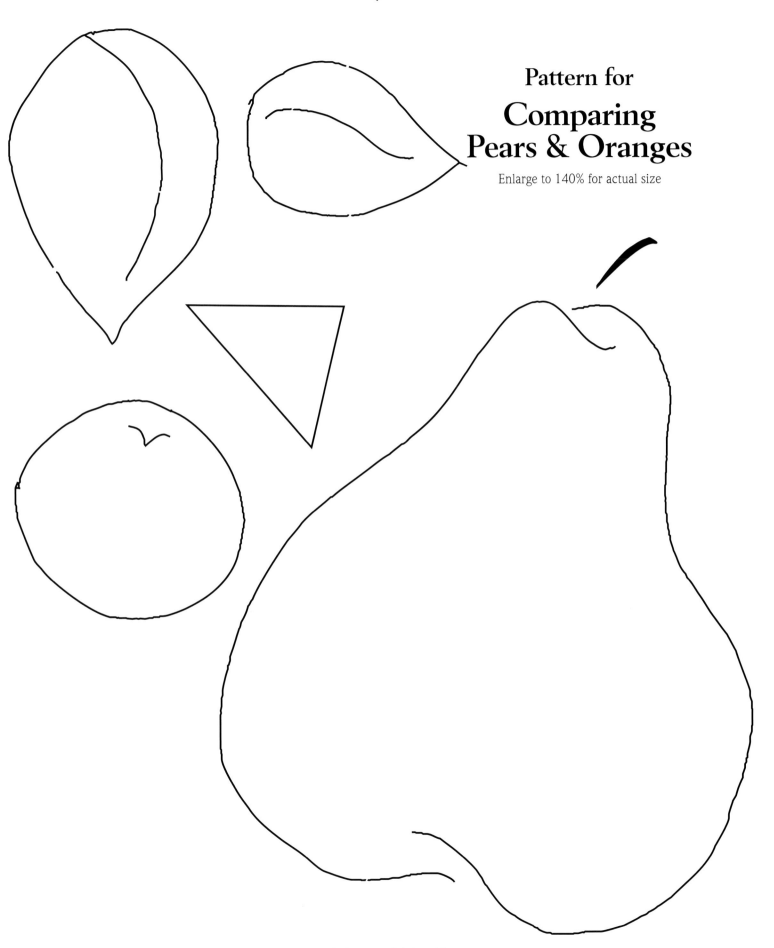

Pattern for
Comparing
Pears & Oranges

Enlarge to 140% for actual size

Harlequin Roses
Painted Canvas Floorcloth

By Patty Cox

A bright blue border with easy to paint roses accents the harlequin diamond shapes in the center of this canvas floorcloth. The canvas used had been dyed the mustard color.

1 • Supplies

Paints & Finishes:

Acrylic craft paint: orange, red, crimson, medium green, light green, magenta, ultramarine, black, creamy yellow
Neutral waterbase glazing medium
polyacrylic semi-gloss protective finish
polyacrylic satin protective finish

Project Surface:

Mustard floorcloth canvas, 56-1/2" x 32-1/2" (finished size: 54-1/2" x 30-1/2")

Tools & Equipment:

3/4" flat brush
#6 liner brush
Wash-away fabric marker
Sponge

2 • Preparation

1. Prepare canvas and turn edges. See "Basic Steps for Making a Canvas Floorcloth."
2. Measure and mark canvas according to Fig. 1, using wash-away fabric marker. If you are not comfortable painting roses free hand, use pattern for Sweetheart Roses on page 99.

3 • Painting

1. Paint orange diamonds, leaving other diamonds mustard base color.
2. Paint roses with an orange wash. Shade with red wash and crimson wash.
3. Paint leaves with a light green wash. Shade with a medium green wash.
4. Paint dark border with ultramarine. Dab some areas with magenta, creating a mottled look.
5. Lightly sponge outer border with creamy yellow. Let dry.
6. Paint a wavy line of orange + light with 3/4" brush.
7. Outline diamonds, borders, roses, and leaves with black, using a liner brush. Let dry.

4 • Finishing

1. Apply 4 coats polyacrylic gloss, allowing to dry between coats.
2. Finish with a coat of polyacrylic satin. ✍

Fig. 1

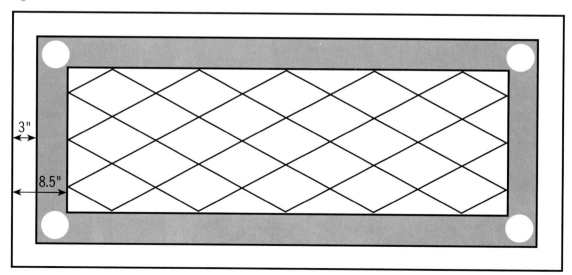

Sweetheart Roses
Painted Canvas Floorcloth

By Patty Cox

Bright and feminine, the floorcloth would be perfect for a little girl's room. All the painting is done with washes of color.

Patterns for this project are on page 99.

1 • Supplies

Paints & Finishes:
Acrylic craft paint: pink, patina, yellow, blue, cranberry, periwinkle
Neutral waterbase glazing medium
Polyacrylic semi-gloss protective finish
Polyacrylic satin protective finish

Project Surface:
White floorcloth canvas, 36-1/2" x 30-1/2" (finished size: 34-1/2" x 28-1/2")

Tools & Equipment:
1/4" flat brush
#6 liner brush
1" sponge brush
Wash-away fabric marker

2 • Preparation

1. Prepare canvas and turn edges. See "Basic Steps for Making a Canvas Floorcloth."
2. Measure and mark canvas according to Fig. 1 using wash-away fabric marker. Pattern for roses and hearts can be transferred to the canvas using the patterns provided, or they can be free-handed using the marker to position them.

3 • Painting

ALL painting on this floorcloth is done with a wash (paint color + water).

1. Paint roses with a 1" sponge brush. Paint a yellow c-stroke in rose center. See Fig. 1 on page 98.
2. Paint scallop petals around center with pink + a little cranberry. See Fig. 2 on page 98. Let dry.
3. Paint cranberry scalloped details lines on petals with a liner brush. See Fig. 3 on page 98.
4. Paint hearts with a pink wash.
5. Outline hearts with cranberry, using a liner brush.
6. Paint leaves with patina, leaving a white highlight on each. See Fig. 3. Outline with blue + periwinkle, using a liner brush. See Fig. 4.
7. Paint background inside scalloped lines with periwinkle + blue, leaving about 1/4" white space around each rose, heart, and leaf.
8. Paint yellow stripes around edge to make a border, using a 1" sponge brush.
9. Paint a pink stripe on each side of each yellow stripes, using a 1/4" flat brush. Let dry.

4 • Finishing

1. Apply 4 coats polyacrylic gloss, allowing to dry between coats.
2. Finish with a coat of polyacrylic satin or matte. ☙

Fig. 1

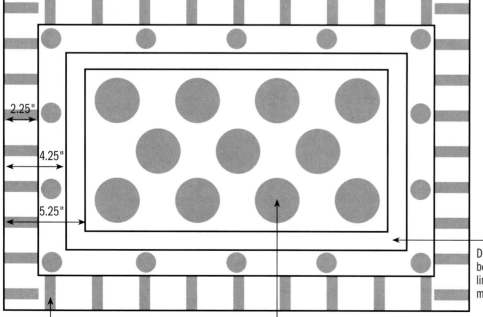

2.25"

4.25"

5.25"

Draw 3" wide scallops between 4.25" and 5.25" lines using a wash-away marker.

Paint 9 yellow 1" border stripes on each side   Paint 11 wash roses in floorcloth center

How to Paint Roses

Fig. 2

Fig. 3

Fig. 4

Fig. 2.
Paint the center of the rose with a wash mix of yellow + water. Use a 1" sponge brush to make a big "C" stroke.

Fig. 3.
Paint rose petals with a pink wash, making big "C" strokes around the center.

Fig. 4.
Paint cranberry petal details using a liner brush.

Pattern for
Sweetheart Rose

Pattern is actual size

Watercolor Roses
Painted Canvas Floorcloth

By Patty Cox

Metallic paints add sparkle and shine to this floorcloth. The design is giving added appeal with a crackle finish. The design is reminiscent of an antique area rug.

1 • Supplies

Paints & Finishes:

Acrylic craft paint: crimson, ultramarine, Prussian blue, aqua, pink, cranberry, parchment, aquamarine, green clover, true teal, metallic antique copper, metallic gold

Neutral waterbase glazing medium

Acrylic crackle medium

Waterbase stain - light oak

Polyacrylic semi-gloss protective finish

Polyacrylic satin protective finish

Project Surface:

Off white floorcloth canvas, 58" x 38" (finished size: 56" x 36")

Tools & Equipment

2" paint brush

#6 liner brush

1" sponge brush

Wash-away fabric marker

continued on page 102

continued from page 100

2 • Preparation

1. Prepare canvas and turn edges. See "Basic Steps for Making a Canvas Floorcloth."
2. Measure and mark canvas according to Fig. 1, using wash-away fabric marker. The roses look much more loose and free if they are simply painted free-hand rather than using a pattern to transfer onto the floorcloth. If you desire, for guidance you may use the wash-away marker to make circles where you will be painting the roses.

3 • Painting

ALL pink and rose painting on this floorcloth is done with a wash (paint color + water).

1. Paint 1" border stripe with pink + cranberry, using a 1" sponge brush.
2. Wash the roses with a 2" sponge brush this way: Paint a center c-stroke of a pink wash. Paint five c-strokes around center with the pink wash. Add some crimson to the wash to make a darker color. Shade each rose with c-strokes. Add additional crimson shading as desired. Allow paint to dry.
3. Paint metallic gold scallops around some of the petals with a liner brush.
4. Paint leaves with a wash of aquamarine + a little green clover, leaving a white highlight on each. Shade with true teal. Allow to dry.

5. Highlight each leaf with a metallic gold stroke, using a liner brush.
6. Paint outer border and oval background with antique copper, leaving about 1/4" white space between border colors and around each flower and leaf. Let dry.
7. Paint crackle medium over antique copper. Let dry.
8. Paint parchment over crackle medium. Allow paint to crackle and dry.
9. Paint rectangular background with ultramarine, leaving 1/4" white space between border and oval. Sponge aqua over paint, lightly sponging outer edges of rectangle.
10. Sponge 1-1/2" area around center oval with Prussian blue.
11. Brush stain on crackled areas. Wipe away excess with rag.
12. Paint metallic gold shadows around roses and leaves. Let dry.

4 • Finishing

1. Apply 4 coats polyacrylic gloss, allowing to dry between coats.
2. Finish with a coat of polyacrylic satin or matte. ✍

Fig. 1

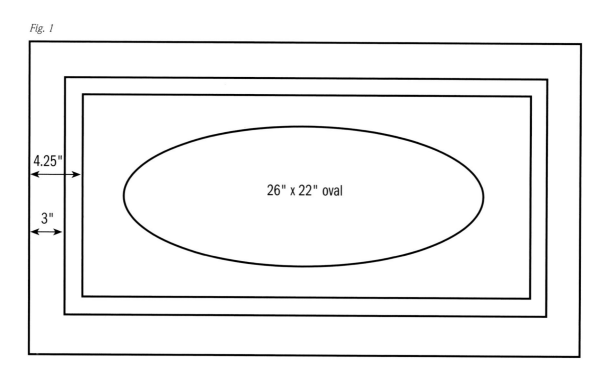

4.25"

3"

26" x 22" oval

How to Paint Roses

Fig. 2

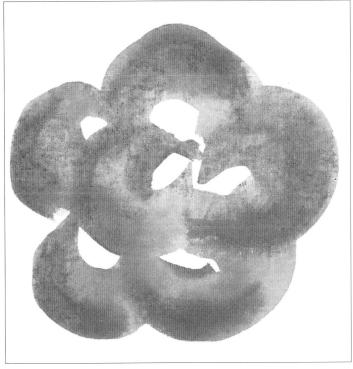

Fig. 3

Fig. 4

Fig. 2
With a 2" sponge brush and a pink wash, paint the center "C" stroke.
Paint "C" stroke petals around the center with the same pink wash.

Fig. 3
Shade with a crimson wash.

Fig. 4
Detail petals with gold scallops

Pastel Leaves
Painted Canvas Floorcloth

By Patty Cox

These summery pastels evoke sunshine all year long. The simple leaf and flower shapes are easy to paint.

Patterns for this project are on page 107.

1 • Supplies

Paints & Finishes:
Acrylic craft paint: chamois, damask blue, green mist, rose, white
Neutral waterbase glazing medium
Polyacrylic semi-gloss protective finish
Polyacrylic satin protective finish

Project Surface:
White floorcloth canvas, 43" x 31" (finished size: 41" x 29")

Tools & Equipment:
1" and 2" sponge brushes
#6 liner brush
Wash-away fabric marker
Sponge

continued on page 106

continued from page 104

2 • Preparation

1. Prepare canvas and turn edges. See "Basic Steps for Making a Canvas Floorcloth."
2. Measure and mark canvas according to Fig. 1, using wash-away fabric marker.
3. To mark the center of the 25" x 17" oval, fold a sheet of newspaper in quarters. Draw a mark 12-1/2" from one fold and 8-1/2" from the other fold. Draw one quarter of the oval. Cut out. Open paper. Place in center of floorcloth. Trace around oval with a wash-away marker.
4. Patterns are provided for the leaf and rosebud. These can be transferred to the floorcloth with transfer paper.

3 • Painting

1. Mix equal amounts rose paint and neutral glazing medium. Brush rose glaze mixture on rose oval border, using a 2" sponge brush.
2. Mix equal amounts chamois + glazing medium. Paint inside oval with chamois glaze mixture. Let dry.

3. Sponge white over oval and oval border. Let dry.
4. Paint circle-shaped roses with rose paint. Let dry. Add a white spiral in each, using a liner brush.
5. Mix equal amounts glazing medium + green mist. Paint one half of each leaf and stems with green glaze mixture. Paint the other half of each leaf with chamois, leaving a white squiggle highlight on each. Allow to dry.
6. Float green mist shading over green side of leaf.
7. Paint a green mist squiggle next to white squiggle highlight on chamois side of leaf.
8. Mix equal amounts damask blue + glazing medium. Paint background with blue glaze mixture, leaving about 1/4" white space around each motif and oval border. Let dry.

4 • Finishing

1. Apply 4 coats polyacrylic gloss, allowing to dry between coats.
2. Finish with a coat of polyacrylic satin or matte. ✎

Fig. 1

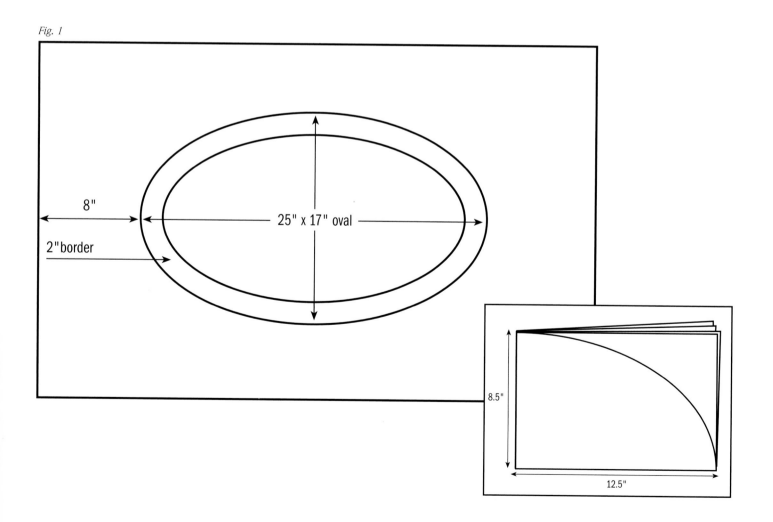

8"

2" border

25" x 17" oval

8.5"

12.5"

Patterns for
Pastel Leaves

Enlarge to 135% for actual size

Tulips & Ivy
Stamped & Stenciled Canvas Floorcloth

By Patty Cox

This floorcloth combines a stamped outer border of pink tulips and a stenciled inner border of green ivy leaves on a yellow lattice background.

1 • Supplies

Paints & Finishes:
Acrylic craft paint: light green, medium green, creamy yellow, warm pink
Neutral waterbase glazing medium
Polyacrylic semi-gloss protective finish
Polyacrylic satin protective finish

Project Surface:
Off white canvas, 32" x 26" (finished size: 30" x 24")

Tools & Equipment:
1" sponge brush
#6 liner brush
Cellulose sponge for making tulip and leaf shapes
Pre-cut ivy stencil
Stencil brush
Wash-away fabric marker
Optional: masking tape

2 • Preparation

1. Prepare canvas and turn edges. See "Basic Steps for Making a Canvas Floorcloth."
2. Measure and mark canvas according to Fig. 1, using a wash-away fabric marker.
3. Using patterns provided, cut tulip and leaf shapes from sponge.

3 • Painting & Stamping

1. Paint eight light green stripes along each long edge, using a 1" sponge brush. Paint six light green stripes along each short edge, leaving a square in each corner.
2. Paint diagonal 1" creamy yellow stripes in rug center to form a lattice design. *Option:* Use tape to mask off the stripes.
3. Dilute creamy yellow paint with water. Paint pale yellow lines between diagonal lines using a #6 liner brush.

4. Dip tulip sponge in warm pink paint. Sponge tulips between each stripe on the border and diagonally in each corner.
5. Dip leaf sponge in medium green paint. Sponge two leaves under each tulip.
6. Paint a thin medium green stripe in the center of each light green border stripe with #6 liner brush.
7. Paint a medium green line on the inside edge of the border with a liner brush. Allow paint to dry.

4 • Stenciling

Stencil ivy with medium green just inside the border, using photo as a guide for placement. Let dry.

5 • Finishing

1. Apply 4 coats polyacrylic gloss, allowing to dry between coats.
2. Finish with a coat of polyacrylic satin. ☙

Fig. 1

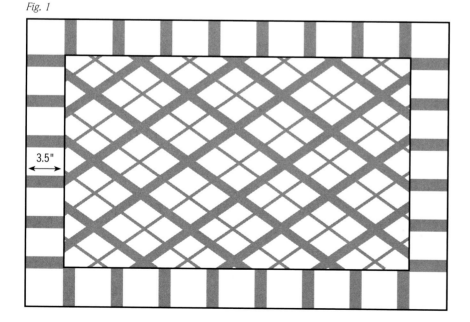

Pattern for tulips (actual size)

For My Pet
Painted Pet's Placemat

By Patty Cox

A personalized placemat is the perfect gift for the pet lover (or your pet). Choose bone shapes for a dog (as shown in the photo) or fish shapes for a favorite feline. Bone appetit!

1 • Supplies

Paints & Finishes:
Acrylic craft paint: creamy yellow, red, aqua, ultramarine, black
Neutral waterbase glazing medium
Polyacrylic semi-gloss protective finish
Polyacrylic satin protective finish

Project Surface:
Off white canvas floorcloth, 20" x 17" (finished size: 18" x 15")

Tools & Equipment:
1" sponge brush
Sponge
1" square sponge
#6 liner brush
Freezer paper

2 • Preparation

1. Prepare canvas and turn edges.
2. Measure and mark canvas according to Fig. 1, using wash-away fabric marker.
3. Cut bones and oval from freezer paper. Iron in position on floorcloth.

3 • Painting

1. Sponge outer area with aqua, sponging over bones.
2. Remove freezer paper. These areas will be white and ready to paint.
3. Mix creamy yellow + water. Paint an oval border 1/4" inside aqua oval, using a 1" sponge brush.
4. Add red to creamy yellow paint to make orange. Paint a 2" orange oval border inside yellow oval border, leaving 1/4" white space between them. Lightly sponge outer edge of orange oval border with red.
5. Sponge 1" squares of ultramarine around outer edge of floorcloth. Use freezer paper pattern to mask bone areas.
6. Outline bone shapes with red, using a liner brush.
7. Print "bone appetit" in corners with wash-away marker. Paint lettering with black.
8. Paint black stripes around yellow oval.
9. Write pet's name in center of floorcloth with wash-away marker. Paint letters with ultramarine, using a liner brush. Let dry.

4 • Finishing

1. Apply 4 coats polyacrylic gloss. allowing to dry between coats.
2. Finish with a coat of polyacrylic satin. ✍

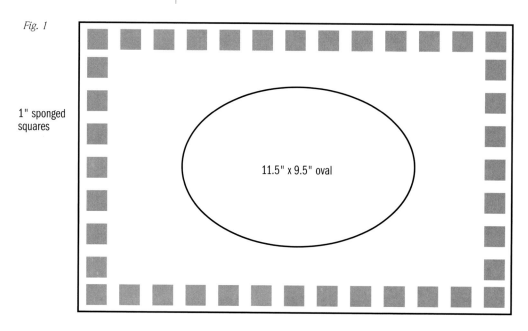

Fig. 1

1" sponged squares

11.5" x 9.5" oval

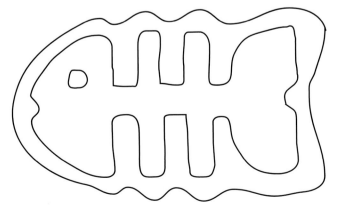

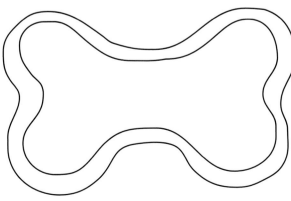

Actual size patterns

Marble at Your Feet
Painted Canvas Floorcloth

By Patty Cox

The marbleizing on this canvas floorcloth was created with sponged veins and brushed streaks of color. The strong geometric shapes enhance the faux marble effect.

1 • Supplies

Paints & Finishes:
Indoor/outdoor acrylic craft paint: black, vanilla, olive, apricot, Mojave sunset
Neutral waterbase glazing medium
Polyacrylic semi-gloss protective finish
Polyacrylic satin protective finish

Project Surface:
Off white floorcloth canvas, 32" x 24" (finished size: 30" x 24")

Tools & Equipment:
Masking tape
2" brush
Sponge
#6 liner brush
Wash-away fabric marker

2 • Preparation

1. Prepare canvas and turn edges. See "Basic Steps for Making a Canvas Floorcloth."
2. Measure and mark canvas according to Fig. 1, using wash-away fabric marker.
3. Tape inside edge of 3" outer border. Apply another row of tape next to that, masking the rug center. Tape around individual black marble squares in outer border.

3 • Painting

Black Areas:
1. Mix equal amounts black paint + glazing medium. Apply to rug center with 2" brush.
2. Paint white streaks over black paint with a #6 liner brush. Hold brush loosely. Twist and drag brush toward you creating uneven, broken lines. Lightly sponge over veins.
3. Lightly sponge olive paint next to some of the white veins.
4. Paint narrow white veins over white sponged veins, twisting and dragging brush to make uneven, broken lines. Paint smaller spider veins in selected areas.
4. Paint all black border squares and center triangles, repeating steps 1-4. Allow paint to dry. Remove tape.

Apricot Areas:
1. Mix equal amounts apricot paint + glazing medium. Apply, using photo as a guide for color placement.
2. Sponge vein pattern with vanilla. Lightly sponge Mojave sunset next to white sponged veins.
3. Mix black + apricot to make gray. Load liner brush with gray. Slide one side of brush into black paint. Twist and drag brush, creating black/gray veins over sponged areas.
4. Paint smaller spider veins with the liner brush in selected areas. Allow paint to dry. Remove tape.

White Areas:
1. Mix equal amounts vanilla paint + glazing medium. Apply, using photo as a guide for placement.

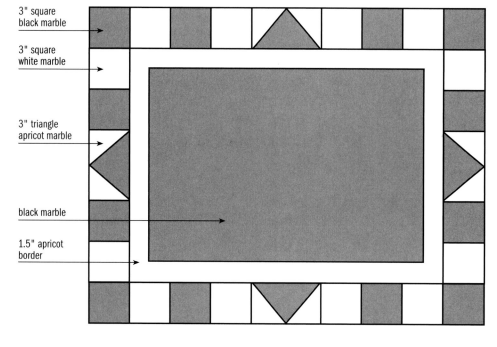

3" square black marble

3" square white marble

3" triangle apricot marble

black marble

1.5" apricot border

Fig. 1

2. Sponge vein pattern with apricot. Using a clean sponge, sponge over apricot to soften color.
3. Lightly sponge black + vanilla vein patterns.
4. Mix black + apricot to make gray. Load liner brush with gray. Slide one side of

brush into black paint. Twist and drag brush creating black/gray veins over sponged areas.
5. Paint smaller spider veins with the liner brush in selected areas. Allow paint to dry. Remove tape.

4 • Finishing

1. Apply 4 coats polyacrylic gloss, allowing to dry between coats.
2. Finish with a coat of polyacrylic satin. ☺

Ivy & Squares
Painted Canvas Floorcloth

By Patty Cox

The soft watercolor effect was created by mixing neutral glazing medium with acrylic paint, then combing the randomly applied colors. The ivy leaf border was stamped.

1 • Supplies

Paints & Finishes:

Acrylic craft paint: Prussian blue, sap green, aqua, antique copper

Neutral waterbase glazing medium

Polyacrylic semi-gloss protective finish

Polyacrylic satin protective finish

Project Surface:

Light green floorcloth canvas, 72" x 42-1/2" (finished size: 70" x 40-1/2")

Tools & Equipment:

3/4" masking tape

2" brush

#10 flat paint brush

Sponge

Pre-cut stamps: ivy leaves

Multi-purpose combing tool

Wash-away fabric marker

Paper towels

2 • Preparation

1. Prepare canvas and turn edges. See "Basic Steps for Making a Canvas Floorcloth."
2. Measure and mark canvas according to Fig. 1, using wash-away fabric marker.
3. Tape inside edge of 3-1/2" outer border.

3 • Painting

Borders:

1. Using a separate container for each color, mix glazing medium with equal amounts of Prussian blue, sap green, and aqua. Generously brush glaze mixtures on 3-1/2" outer border, working a 2-3 foot strip at a time, to create a mottled effect.
2. Pull comb over wet glazes in a wave motion from the inside edge of border to outer edge. Wipe glaze off comb with a paper towel.

3. Repeat to complete 3-1/2" border. Reserve remaining glaze mixtures.
4. Tape edges of 1-3/4" border. Lightly sponge border with aqua. Remove tape.

Squares:

1. Mask off eight outer dark squares with masking tape. Brush squares with Prussian blue, sap green and aqua glaze mixtures.

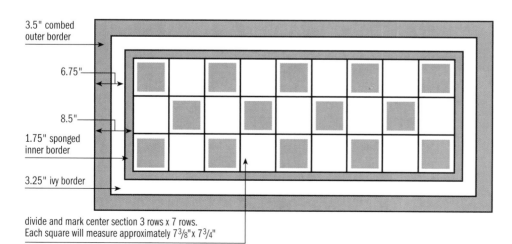

3.5" combed outer border

6.75"

8.5"

1.75" sponged inner border

3.25" ivy border

divide and mark center section 3 rows x 7 rows.
Each square will measure approximately 7⅜"x 7¾"

Fig. 1

2. While glazes are wet, pull comb lengthwise over each square. Wipe glaze off comb with a paper towel. Repeat for all eight squares. Let dry. Remove tape.

3. Mask off three inner dark squares. Glaze and comb these squares, using the same colors.

4. Mask off unpainted squares between combed squares. Lightly sponge aqua around edges of each square. Let dry. Remove tape.

4 • Stamping

Load ivy leaf stamps by applying paint to stamp with a flat brush. Use two to three paint colors on each, highlighting some edges with copper. Stamp ivy leaves around unpainted border, using photo as a guide. Begin with the large leaves at the center of each side, trailing to smallest leaves in corners. Let dry.

5 • Finishing

1. Apply 4 coats polyacrylic gloss, allowing to dry between coats.

2. Finish with a coat of polyacrylic satin. 🖎

Tea Towel
Stamped Canvas Floorcloth

By Patty Cox

An antique tea towel was the inspiration for this floorcloth. The stamped fruits and leaves are easy to do, and the colors will brighten up your kitchen!

1 • Supplies

Paints & Finishes:

Colored paint glazes: lemon yellow, persimmon, sunflower, apricot, plum, deep purple, leaf green, ivy green

Acrylic craft paint: white

Neutral waterbase glazing medium

Polyacrylic semi-gloss protective finish

Polyacrylic satin protective finish

Project Surface:

White floorcloth canvas, 30" x 26" (finished size: 28" x 24")

Tools & Equipment

1" sponge brush

1/2" acrylic brush

#6 liner brush

Stippler brush

Wash-away fabric marker

Pre-cut stamps: peach, pear

2 round paper doilies, 8" diameter

Spray adhesive

2 • Preparation

1. Prepare canvas and turn edges. See "Basic Steps for Making a Canvas Floorcloth."
2. Measure and mark canvas according to Fig. 1, using wash-away fabric marker.

3 • Painting

1. Mix equal amounts persimmon glaze and glazing medium. Dilute with water. Paint red stripes to make border.
2. Mix equal amounts lemon yellow glaze and glazing medium. Dilute with water. Paint yellow stripes to complete border.
3. Cut 2" off sides of paper doily. Center 2" sections along outer yellow stripe of border. Trace around doily (just the outlines) with wash-away marker. Paint inside traced lines with lemon yellow. Let dry.
4. Apply spray adhesive to back of each doily section. Press doily over yellow painted area. Using white acrylic paint and stipple brush, stencil doily pattern. Let dry.

4 • Stamping

Stamp fruit and leaves over painted stripes, mixing glazing medium with colored glazes on stamps. Use the peach stamp to make plums as well as peaches. Here are the color combinations:

Pears — sunflower + persimmon, shaded with persimmon.

Peaches — apricot + persimmon, shaded with persimmon.

Plums — plum + deep purple, shaded with deep purple.

Leaves — lemon yellow + leaf green, shaded with leaf green OR ivy green + leaf green, shaded with ivy green. Let dry.

5 • Finishing

1. Apply 4 coats polyacrylic gloss, allowing to dry between coats.
2. Finish with a coat of polyacrylic satin. ✍

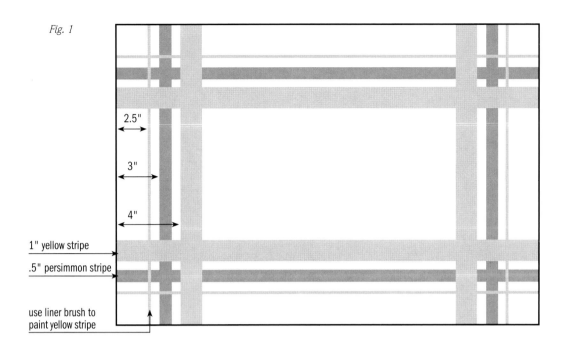

Fig. 1

2.5"

3"

4"

1" yellow stripe

.5" persimmon stripe

use liner brush to
paint yellow stripe

PATTERNS

Pattern for
Hexagon & Diamonds

Instructions on page 26
Enlarge to 190% for actual size

Use the shaded area to make additional overlays for darker stenciling

Pattern for
Leaf Garland

Instructions on page 28
Enlarge to 105% for actual size

Cut a separate stencil overlay
for the stem/vein section

Pattern for
Fabulous Ferns

Instructions on page 32
Enlarge to 265%

Pattern for
La Richesse Border

Instructions on page 44
Enlarge to 152%

Pattern for
Compass Points

Instructions on page 54
Actual size pattern

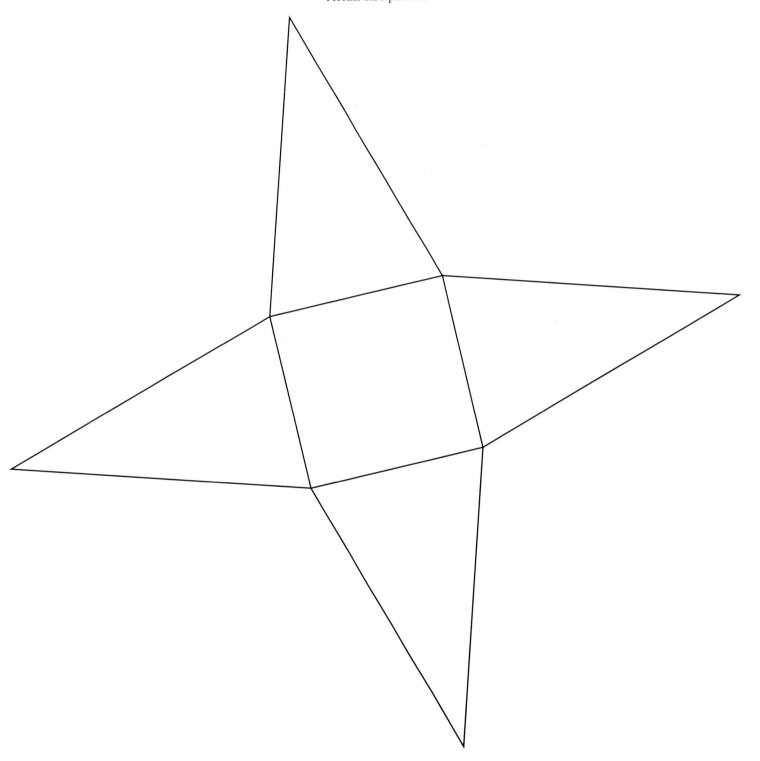

Pattern for
Sisal Beauty

Instructions on page 74
Enlarge to 276% for actual size

Shaded area
is for white
overlay

Pattern for
Magnolia Elegance

Instructions on page 78
Enlarge to 330% for actual size

Pattern for
Harvested Fruit

Instructions on page 80
Enlarge to 255% for actual size

Pattern for
Medallion Floorcloth & Patio Medallions

Instructions on page 84
Enlarge 135% for
actual size

Pattern for
Tulips & Squares

Instructions on page 40
Enlarge to 135% for actual size

METRIC CONVERSION CHART

Inches to Millimeters and Centimeters

Inches	MM	CM
1/8	3	.3
1/4	6	.6
3/8	10	1.0
1/2	13	1.3
5/8	16	1.6
3/4	19	1.9
7/8	22	2.2
1	25	2.5
1-1/4	32	3.2
1-1/2	38	3.8
1-3/4	44	4.4
2	51	5.1
3	76	7.6
4	102	10.2
5	127	12.7
6	152	15.2
7	178	17.8
8	203	20.3
9	229	22.9
10	254	25.4
11	279	27.9
12	305	30.5

Yards to Meters

Yards	Meters
1/8	.11
1/4	.23
3/8	.34
1/2	.46
5/8	.57
3/4	.69
7/8	.80
1	.91
2	1.83
3	2.74
4	3.66
5	4.57
6	5.49
7	6.40
8	7.32
9	8.23
10	9.14

INDEX